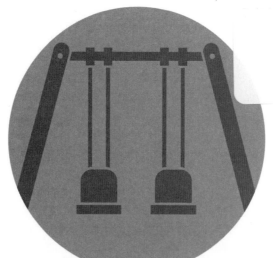

Swing Set Makeover

STEM Road Map
for Elementary School

Grade
3

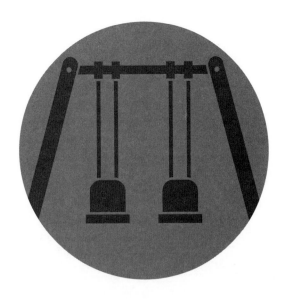

Swing Set Makeover

Grade
3

STEM Road Map
for Elementary School

Edited by Carla C. Johnson, Janet B. Walton, and
Erin Peters-Burton

National Science Teachers Association

Arlington, Virginia

National Science Teachers Association

Claire Reinburg, Director
Rachel Ledbetter, Managing Editor
Deborah Siegel, Associate Editor
Andrea Silen, Associate Editor
Donna Yudkin, Book Acquisitions Manager

ART AND DESIGN
Will Thomas Jr., Director, cover and
    interior design
Himabindu Bichali, Graphic Designer, interior
    design

PRINTING AND PRODUCTION
Catherine Lorrain, Director

NATIONAL SCIENCE TEACHERS ASSOCIATION
David L. Evans, Executive Director

1840 Wilson Blvd., Arlington, VA 22201
www.nsta.org/store
For customer service inquiries, please call 800-277-5300.

NSTA is committed to publishing material that promotes the best in inquiry-based science education. However, conditions of actual use may vary, and the safety procedures and practices described in this book are intended to serve only as a guide. Additional precautionary measures may be required. NSTA and the authors do not warrant or represent that the procedures and practices in this book meet any safety code or standard of federal, state, or local regulations. NSTA and the authors disclaim any liability for personal injury or damage to property arising out of or relating to the use of this book, including any of the recommendations, instructions, or materials contained therein.

Library of Congress Cataloging-in-Publication Data
Names: Johnson, Carla C., 1969- editor. | Walton, Janet B., 1968- editor. | Peters-Burton, Erin E., editor.
Title: Swing set makeover, grade 3 : STEM road map for elementary school / Edited by Carla C. Johnson,
    Janet B. Walton, and Erin Peters-Burton.
Other titles: Swing set makeover, grade three
Description: Arlington, VA : National Science Teachers Association, [2018] | Includes bibliographical references
    and index.
Identifiers: LCCN 2018020161 (print) | LCCN 2018029109 (ebook) | ISBN 9781681404639 (e-book) |
    ISBN 9781681404622 (print)
Subjects: LCSH: Swings--Design and construction. | Science--Study and teaching (Elementary) |
    Engineering--Study and teaching (Elementary)
Classification: LCC TT176 (ebook) | LCC TT176 .S94 2018 (print) | DDC 684.1/8--dc23
LC record available at *https://lccn.loc.gov/2018020161*

The *Next Generation Science Standards* ("NGSS") were developed by twenty-six states, in collaboration with the National Research Council, the National Science Teachers Association and the American Association for the Advancement of Science in a process managed by Achieve, Inc. For more information go to *www.nextgenscience.org*.

CONTENTS

CONTENTS

ABOUT THE EDITORS AND AUTHORS

Dr. Carla C. Johnson is the associate dean for research, engagement, and global partnerships and a professor of science education at Purdue University's College of Education in West Lafayette, Indiana. Dr. Johnson serves as the director of research and evaluation for the Department of Defense–funded Army Educational Outreach Program (AEOP), a global portfolio of STEM education programs, competitions, and apprenticeships. She has been a leader in STEM education for the past decade, serving as the director of STEM Centers, editor of the *School Science and Mathematics* journal, and lead researcher for the evaluation of Tennessee's Race to the Top–funded STEM portfolio. Dr. Johnson has published over 100 articles, books, book chapters, and curriculum books focused on STEM education. She is a former science and social studies teacher and was the recipient of the 2013 Outstanding Science Teacher Educator of the Year award from the Association for Science Teacher Education (ASTE), the 2012 Award for Excellence in Integrating Science and Mathematics from the School Science and Mathematics Association (SSMA), the 2014 award for best paper on Implications of Research for Educational Practice from ASTE, and the 2006 Outstanding Early Career Scholar Award from SSMA. Her research focuses on STEM education policy implementation, effective science teaching, and integrated STEM approaches.

Dr. Janet B. Walton is a research assistant professor and the assistant director of evaluation for AEOP at Purdue University's College of Education. Formerly the STEM workforce program manager for Virginia's Region 2000 and founding director of the Future Focus Foundation, a nonprofit organization dedicated to enhancing the quality of STEM education in the region, she merges her economic development and education backgrounds to develop K–12 curricular materials that integrate real-life issues with sound cross-curricular content. Her research focuses on collaboration between schools and community stakeholders for STEM education and problem- and project-based learning pedagogies. With this research agenda, she works to forge productive relationships between K–12 schools and local business and community stakeholders to bring contextual STEM experiences into the classroom and provide students and educators with innovative resources and curricular materials.

Dr. Erin Peters-Burton is the Donna R. and David E. Sterling endowed professor in science education at George Mason University in Fairfax, Virginia. She uses her experiences from 15 years as an engineer and secondary science, engineering, and mathematics teacher to develop research projects that directly inform classroom practice in science and engineering. Her research agenda is based on the idea that all students should build self-awareness of how they learn science and engineering. She works to help students see themselves as "science-minded" and help teachers create classrooms that support student skills to develop scientific knowledge. To accomplish this, she pursues research projects that investigate ways that students and teachers can use self-regulated learning theory in science and engineering, as well as how inclusive STEM schools can help students succeed. During her tenure as a secondary teacher, she had a National Board Certification in Early Adolescent Science and was an Albert Einstein Distinguished Educator Fellow for NASA. As a researcher, Dr. Peters-Burton has published over 100 articles, books, book chapters, and curriculum books focused on STEM education and educational psychology. She received the Outstanding Science Teacher Educator of the Year award from ASTE in 2016 and a Teacher of Distinction Award and a Scholarly Achievement Award from George Mason University in 2012, and in 2010 she was named University Science Educator of the Year by the Virginia Association of Science Teachers.

Dr. Tamara J. Moore is an associate professor of engineering education in the College of Engineering at Purdue University. Dr. Moore's research focuses on defining STEM integration through the use of engineering as the connection and investigating its power for student learning.

Paula Schoeff taught grades K–8 for 20 years in Indiana and Ohio and assisted in helping educate teachers in Ohio and Kentucky in STEM practices. Schoeff has a master's degree from the University of Cincinnati in curriculum and instruction with a focus on science.

Dr. Toni A. Sondergeld is an associate professor of assessment, research, and statistics in the School of Education at Drexel University in Philadelphia. Dr. Sondergeld's research concentrates on assessment and evaluation in education, with a focus on K–12 STEM.

ACKNOWLEDGMENTS

This module was developed as a part of the STEM Road Map project (Carla C. Johnson, principal investigator). The Purdue University College of Education, General Motors, and other sources provided funding for this project.

PART 1

THE STEM ROAD MAP

BACKGROUND, THEORY, AND PRACTICE

OVERVIEW OF THE *STEM ROAD MAP CURRICULUM SERIES*

Carla C. Johnson, Erin Peters-Burton, and Tamara J. Moore

The *STEM Road Map Curriculum Series* was conceptualized and developed by a team of STEM educators from across the United States in response to a growing need to infuse real-world learning contexts, delivered through authentic problem-solving pedagogy, into K–12 classrooms. The curriculum series is grounded in integrated STEM, which focuses on the integration of the STEM disciplines—science, technology, engineering, and mathematics—delivered across content areas, incorporating the Framework for 21st Century Learning along with grade-level-appropriate academic standards.

The curriculum series begins in kindergarten, with a five-week instructional sequence that introduces students to the STEM themes and gives them grade-level-appropriate topics and real-world challenges or problems to solve. The series uses project-based and problem-based learning, presenting students with the problem or challenge during the first lesson, and then teaching them science, social studies, English language arts, mathematics, and other content, as they apply what they learn to the challenge or problem at hand.

Authentic assessment and differentiation are embedded throughout the modules. Each *STEM Road Map Curriculum Series* module has a lead discipline, which may be science, social studies, English language arts, or mathematics. All disciplines are integrated into each module, along with ties to engineering. Another key component is the use of STEM Research Notebooks to allow students to track their own learning progress. The modules are designed with a scaffolded approach, with increasingly complex concepts and skills introduced as students progress through grade levels.

The developers of this work view the curriculum as a resource that is intended to be used either as a whole or in part to meet the needs of districts, schools, and teachers who are implementing an integrated STEM approach. A variety of implementation formats are possible, from using one stand-alone module at a given grade level to using all five modules to provide 25 weeks of instruction. Also, within each grade band (K–2, 3–5, 6–8, 9–12), the modules can be sequenced in various ways to suit specific needs.

STANDARDS-BASED APPROACH

The *STEM Road Map Curriculum Series* is anchored in the *Next Generation Science Standards (NGSS)*, the *Common Core State Standards for Mathematics (CCSS Mathematics)*, the *Common Core State Standards for English Language Arts (CCSS ELA)*, and the Framework for 21st Century Learning. Each module includes a detailed curriculum map that incorporates the associated standards from the particular area correlated to lesson plans. The STEM Road Map has very clear and strong connections to these academic standards, and each of the grade-level topics was derived from the mapping of the standards to ensure alignment among topics, challenges or problems, and the required academic standards for students. Therefore, the curriculum series takes a standards-based approach and is designed to provide authentic contexts for application of required knowledge and skills.

THEMES IN THE *STEM ROAD MAP CURRICULUM SERIES*

The K–12 STEM Road Map is organized around five real-world STEM themes that were generated through an examination of the big ideas and challenges for society included in STEM standards and those that are persistent dilemmas for current and future generations:

- Cause and Effect

- Innovation and Progress

- The Represented World

- Sustainable Systems

- Optimizing the Human Experience

These themes are designed as springboards for launching students into an exploration of real-world learning situated within big ideas. Most important, the five STEM Road Map themes serve as a framework for scaffolding STEM learning across the K–12 continuum.

The themes are distributed across the STEM disciplines so that they represent the big ideas in science (Cause and Effect; Sustainable Systems), technology (Innovation and Progress; Optimizing the Human Experience), engineering (Innovation and Progress; Sustainable Systems; Optimizing the Human Experience), and mathematics (The Represented World), as well as concepts and challenges in social studies and 21st century skills that are also excellent contexts for learning in English language arts. The process of developing themes began with the clustering of the *NGSS* performance expectations and the National Academy of Engineering's grand challenges for engineering, which led to the development of the challenge in each module and connections of the module activities to the *CCSS Mathematics* and *CCSS ELA* standards. We performed these mapping processes with large teams of experts and found that these five themes provided breadth, depth, and coherence to frame a high-quality STEM learning experience from kindergarten through 12th grade.

Cause and Effect

The concept of cause and effect is a powerful and pervasive notion in the STEM fields. It is the foundation of understanding how and why things happen as they do. Humans spend considerable effort and resources trying to understand the causes and effects of natural and designed phenomena to gain better control over events and the environment and to be prepared to react appropriately. Equipped with the knowledge of a specific cause-and-effect relationship, we can lead better lives or contribute to the community by altering the cause, leading to a different effect. For example, if a person recognizes that irresponsible energy consumption leads to global climate change, that person can act to remedy his or her contribution to the situation. Although cause and effect is a core idea in the STEM fields, it can actually be difficult to determine. Students should be capable of understanding not only when evidence points to cause and effect but also when evidence points to relationships but not direct causality. The major goal of education is to foster students to be empowered, analytic thinkers, capable of thinking through complex processes to make important decisions. Understanding causality, as well as when it cannot be determined, will help students become better consumers, global citizens, and community members.

Innovation and Progress

One of the most important factors in determining whether humans will have a positive future is innovation. Innovation is the driving force behind progress, which helps create possibilities that did not exist before. Innovation and progress are creative entities, but in the STEM fields, they are anchored by evidence and logic, and they use established concepts to move the STEM fields forward. In creating something new, students must consider what is already known in the STEM fields and apply this knowledge appropriately. When we innovate, we create value that was not there previously and create new conditions and possibilities for even more innovations. Students should consider how their innovations might affect progress and use their STEM thinking to change current human burdens to benefits. For example, if we develop more efficient cars that use byproducts from another manufacturing industry, such as food processing, then we have used waste productively and reduced the need for the waste to be hauled away, an indirect benefit of the innovation.

The Represented World

When we communicate about the world we live in, how the world works, and how we can meet the needs of humans, sometimes we can use the actual phenomena to explain a concept. Sometimes, however, the concept is too big, too slow, too small, too fast, or too complex for us to explain using the actual phenomena, and we must use a representation or a model to help communicate the important features. We need representations and models such as graphs, tables, mathematical expressions, and diagrams because it makes our thinking visible. For example, when examining geologic time, we cannot actually observe the passage of such large chunks of time, so we create a timeline or a model that uses a proportional scale to visually illustrate how much time has passed for different eras. Another example may be something too complex for students at a particular grade level, such as explaining the *p* subshell orbitals of electrons to fifth graders. Instead, we use the Bohr model, which more closely represents the orbiting of planets and is accessible to fifth graders.

When we create models, they are helpful because they point out the most important features of a phenomenon. We also create representations of the world with mathematical functions, which help us change parameters to suit the situation. Creating representations of a phenomenon engages students because they are able to identify the important features of that phenomenon and communicate them directly. But because models are estimates of a phenomenon, they leave out some of the details, so it is important for students to evaluate their usefulness as well as their shortcomings.

Sustainable Systems

From an engineering perspective, the term *system* refers to the use of "concepts of component need, component interaction, systems interaction, and feedback. The interaction of subcomponents to produce a functional system is a common lens used by all engineering disciplines for understanding, analysis, and design." (Koehler, Bloom, and Binns 2013, p. 8). Systems can be either open (e.g., an ecosystem) or closed (e. g., a car battery). Ideally, a system should be sustainable, able to maintain equilibrium without much energy from outside the structure. Looking at a garden, we see flowers blooming, weeds sprouting, insects buzzing, and various forms of life living within its boundaries. This is an example of an ecosystem, a collection of living organisms that survive together, functioning as a system. The interaction of the organisms within the system and the influences of the environment (e.g., water, sunlight) can maintain the system for a period of time, thus demonstrating its ability to endure. Sustainability is a desirable feature of a system because it allows for existence of the entity in the long term.

In the STEM Road Map project, we identified different standards that we consider to be oriented toward systems that students should know and understand in the K–12 setting. These include ecosystems, the rock cycle, Earth processes (such as erosion, tectonics, ocean currents, weather phenomena), Earth-Sun-Moon cycles, heat transfer, and the interaction among the geosphere, biosphere, hydrosphere, and atmosphere. Students and teachers should understand that we live in a world of systems that are not independent of each other, but rather are intrinsically linked such that a disruption in one part of a system will have reverberating effects on other parts of the system.

Optimizing the Human Experience

Science, technology, engineering, and mathematics as disciplines have the capacity to continuously improve the ways humans live, interact, and find meaning in the world, thus working to optimize the human experience. This idea has two components: being more suited to our environment and being more fully human. For example, the progression of STEM ideas can help humans create solutions to complex problems, such as improving ways to access water sources, designing energy sources with minimal impact on our environment, developing new ways of communication and expression, and building efficient shelters. STEM ideas can also provide access to the secrets and wonders of nature. Learning in STEM requires students to think logically and systematically, which is a way of knowing the world that is markedly different from knowing the world as an artist. When students can employ various ways of knowing and understand when it is appropriate to use a different way of knowing or integrate ways of knowing, they are fully experiencing the best of what it is to be human. The problem-based learning scenarios provided in the STEM Road Map help students develop ways of thinking like STEM professionals as they ask questions and design solutions. They learn to optimize the human experience by innovating improvements in the designed world in which they live.

THE NEED FOR AN INTEGRATED STEM APPROACH

At a basic level, STEM stands for science, technology, engineering, and mathematics. Over the past decade, however, STEM has evolved to have a much broader scope and broader implications. Now, educators and policy makers refer to STEM as not only a concentrated area for investing in the future of the United States and other nations but also as a domain and mechanism for educational reform.

The good intentions of the recent decade-plus of focus on accountability and increased testing has resulted in significant decreases not only in instructional time for teaching science and social studies but also in the flexibility of teachers to promote authentic, problem solving–focused classroom environments. The shift has had a detrimental impact on student acquisition of vitally important skills, which many refer to as 21st century skills, and often the ability of students to "think." Further, schooling has become increasingly siloed into compartments of mathematics, science, English language arts, and social studies, lacking any of the connections that are overwhelmingly present in

the real world around children. Students have experienced school as content provided in boxes that must be memorized, devoid of any real-world context, and often have little understanding of why they are learning these things.

STEM-focused projects, curriculum, activities, and schools have emerged as a means to address these challenges. However, most of these efforts have continued to focus on the individual STEM disciplines (predominantly science and engineering) through more STEM classes and after-school programs in a "STEM enhanced" approach (Breiner et al. 2012). But in traditional and STEM enhanced approaches, there is little to no focus on other disciplines that are integral to the context of STEM in the real world. Integrated STEM education, on the other hand, infuses the learning of important STEM content and concepts with a much-needed emphasis on 21st century skills and a problem- and project-based pedagogy that more closely mirrors the real-world setting for society's challenges. It incorporates social studies, English language arts, and the arts as pivotal and necessary (Johnson 2013; Rennie, Venville, and Wallace 2012; Roehrig et al. 2012).

FRAMEWORK FOR STEM INTEGRATION IN THE CLASSROOM

The *STEM Road Map Curriculum Series* is grounded in the Framework for STEM Integration in the Classroom as conceptualized by Moore, Guzey, and Brown (2014) and Moore et al. (2014). The framework has six elements, described in the context of how they are used in the *STEM Road Map Curriculum Series* as follows:

1. The STEM Road Map contexts are meaningful to students and provide motivation to engage with the content. Together, these allow students to have different ways to enter into the challenge.

2. The STEM Road Map modules include engineering design that allows students to design technologies (i.e., products that are part of the designed world) for a compelling purpose.

3. The STEM Road Map modules provide students with the opportunities to learn from failure and redesign based on the lessons learned.

4. The STEM Road Map modules include standards-based disciplinary content as the learning objectives.

5. The STEM Road Map modules include student-centered pedagogies that allow students to grapple with the content, tie their ideas to the context, and learn to think for themselves as they deepen their conceptual knowledge.

6. The STEM Road Map modules emphasize 21st century skills and, in particular, highlight communication and teamwork.

All of the STEM Road Map modules incorporate these six elements; however, the level of emphasis on each of these elements varies based on the challenge or problem in each module.

THE NEED FOR THE *STEM ROAD MAP CURRICULUM SERIES*

As focus is increasing on integrated STEM, and additional schools and programs decide to move their curriculum and instruction in this direction, there is a need for high-quality, research-based curriculum designed with integrated STEM at the core. Several good resources are available to help teachers infuse engineering or more STEM enhanced approaches, but no curriculum exists that spans K–12 with an integrated STEM focus. The next chapter provides detailed information about the specific pedagogy, instructional strategies, and learning theory on which the *STEM Road Map Curriculum Series* is grounded.

REFERENCES

Breiner, J., M. Harkness, C. C. Johnson, and C. Koehler. 2012. What is STEM? A discussion about conceptions of STEM in education and partnerships. *School Science and Mathematics* 112 (1): 3–11.

Johnson, C. C. 2013. Conceptualizing integrated STEM education: Editorial. *School Science and Mathematics* 113 (8): 367–368.

Koehler, C. M., M. A. Bloom, and I. C. Binns. 2013. Lights, camera, action: Developing a methodology to document mainstream films' portrayal of nature of science and scientific inquiry. *Electronic Journal of Science Education* 17 (2).

Moore, T. J., S. S. Guzey, and A. Brown. 2014. Greenhouse design to increase habitable land: An engineering unit. *Science Scope* 51–57.

Moore, T. J., M. S. Stohlmann, H.-H. Wang, K. M. Tank, A. W. Glancy, and G. H. Roehrig. 2014. Implementation and integration of engineering in K–12 STEM education. In *Engineering in pre-college settings: Synthesizing research, policy, and practices,* ed. S. Purzer, J. Strobel, and M. Cardella, 35–60. West Lafayette, IN: Purdue Press.

Rennie, L., G. Venville, and J. Wallace. 2012. *Integrating science, technology, engineering, and mathematics: Issues, reflections, and ways forward.* New York: Routledge.

Roehrig, G. H., T. J. Moore, H. H. Wang, and M. S. Park. 2012. Is adding the *E* enough? Investigating the impact of K–12 engineering standards on the implementation of STEM integration. *School Science and Mathematics* 112 (1): 31–44.

STRATEGIES USED IN THE *STEM ROAD MAP CURRICULUM SERIES*

Erin Peters-Burton, Carla C. Johnson, Toni A. Sondergeld, and Tamara J. Moore

The *STEM Road Map Curriculum Series* uses what has been identified through research as best-practice pedagogy, including embedded formative assessment strategies throughout each module. This chapter briefly describes the key strategies that are employed in the series.

PROJECT- AND PROBLEM-BASED LEARNING

Each module in the *STEM Road Map Curriculum Series* uses either project-based learning or problem-based learning to drive the instruction. Project-based learning begins with a driving question to guide student teams in addressing a contextualized local or community problem or issue. The outcome of project-based instruction is a product that is conceptualized, designed, and tested through a series of scaffolded learning experiences (Blumenfeld et al. 1991; Krajcik and Blumenfeld 2006). Problem-based learning is often grounded in a fictitious scenario, challenge, or problem (Barell 2006; Lambros 2004). On the first day of instruction within the unit, student teams are provided with the context of the problem. Teams work through a series of activities and use open-ended research to develop their potential solution to the problem or challenge, which need not be a tangible product (Johnson 2003).

ENGINEERING DESIGN PROCESS

The *STEM Road Map Curriculum Series* uses engineering design as a way to facilitate integrated STEM within the modules. The engineering design process (EDP) is depicted in Figure 2.1 (p. 10). It highlights two major aspects of engineering design—problem scoping and solution generation—and six specific components of working toward a design: define the problem, learn about the problem, plan a solution, try the solution, test the solution, decide whether the solution is good enough. It also shows that communication

Figure 2.1. Engineering Design Process

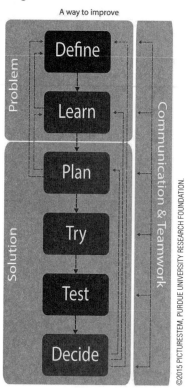

and teamwork are involved throughout the entire process. As the arrows in the figure indicate, the order in which the components of engineering design are addressed depends on what becomes needed as designers progress through the EDP. Designers must communicate and work in teams throughout the process. The EDP is iterative, meaning that components of the process can be repeated as needed until the design is good enough to present to the client as a potential solution to the problem.

Problem scoping is the process of gathering and analyzing information to deeply understand the engineering design problem. It includes defining the problem and learning about the problem. Defining the problem includes identifying the problem, the client, and the end user of the design. The client is the person (or people) who hired the designers to do the work, and the end user is the person (or people) who will use the final design. The designers must also identify the criteria and the constraints of the problem. The criteria are the things the client wants from the solution, and the constraints are the things that limit the possible solutions. The designers must spend significant time learning about the problem, which can include activities such as the following:

- Reading informational texts and researching about relevant concepts or contexts

- Identifying and learning about needed mathematical and scientific skills, knowledge, and tools

- Learning about things done previously to solve similar problems

- Experimenting with possible materials that could be used in the design

Problem scoping also allows designers to consider how to measure the success of the design in addressing specific criteria and staying within the constraints over multiple iterations of solution generation.

Solution generation includes planning a solution, trying the solution, testing the solution, and deciding whether the solution is good enough. Planning the solution includes generating many design ideas that both address the criteria and meet the constraints. Here the designers must consider what was learned about the problem during problem scoping. Design plans include clear communication of design ideas through media such as notebooks, blueprints, schematics, or storyboards. They also include details about the design, such as measurements, materials, colors, costs of materials, instructions for how things fit together, and sets of directions. Making the decision about which design idea to move forward involves considering the trade-offs of each design idea.

Once a clear design plan is in place, the designers must try the solution. Trying the solution includes developing a prototype (a testable model) based on the plan generated. The prototype might be something physical or a process to accomplish a goal. This component of design requires that the designers consider the risk involved in implementing the design. The prototype developed must be tested. Testing the solution includes conducting fair tests that verify whether the plan is a solution that is good enough to meet the client and end user needs and wants. Data need to be collected about the results of the tests of the prototype, and these data should be used to make evidence-based decisions regarding the design choices made in the plan. Here, the designers must again consider the criteria and constraints for the problem.

Using the data gathered from the testing, the designers must decide whether the solution is good enough to meet the client and end user needs and wants by assessment based on the criteria and constraints. Here, the designers must justify or reject design decisions based on the background research gathered while learning about the problem and on the evidence gathered during the testing of the solution. The designers must now decide whether to present the current solution to the client as a possibility or to do more iterations of design on the solution. If they decide that improvements need to be made to the solution, the designers must decide if there is more that needs to be understood about the problem, client, or end user; if another design idea should be tried; or if more planning needs to be conducted on the same design. One way or another, more work needs to be done.

Throughout the process of designing a solution to meet a client's needs and wants, designers work in teams and must communicate to each other, the client, and likely the end user. Teamwork is important in engineering design because multiple perspectives and differing skills and knowledge are valuable when working to solve problems. Communication is key to the success of the designed solution. Designers must communicate their ideas clearly using many different representations, such as text in an engineering notebook, diagrams, flowcharts, technical briefs, or memos to the client.

LEARNING CYCLE

The same format for the learning cycle is used in all grade levels throughout the STEM Road Map, so that students engage in a variety of activities to learn about phenomena in the modules thoroughly and have consistent experiences in the problem- and project-based learning modules. Expectations for learning by younger students are not as high as for older students, but the format of the progression of learning is the same. Students who have learned with curriculum from the STEM Road Map in early grades know

what to expect in later grades. The learning cycle consists of five parts—Introductory Activity/Engagement, Activity/Exploration, Explanation, Elaboration/Application of Knowledge, and Evaluation/Assessment—and is based on the empirically tested 5E model from BSCS (Bybee et al. 2006).

In the Introductory Activity/Engagement phase, teachers introduce the module challenge and use a unique approach designed to pique students' curiosity. This phase gets students to start thinking about what they already know about the topic and begin wondering about key ideas. The Introductory Activity/Engagement phase positions students to be confident about what they are about to learn, because they have prior knowledge, and clues them into what they don't yet know.

In the Activity/Exploration phase, the teacher sets up activities in which students experience a deeper look at the topics that were introduced earlier. Students engage in the activities and generate new questions or consider possibilities using preliminary investigations. Students work independently, in small groups, and in whole-group settings to conduct investigations, resulting in common experiences about the topic and skills involved in the real-world activities. Teachers can assess students' development of concepts and skills based on the common experiences during this phase.

During the Explanation phase, teachers direct students' attention to concepts they need to understand and skills they need to possess to accomplish the challenge. Students participate in activities to demonstrate their knowledge and skills to this point, and teachers can pinpoint gaps in student knowledge during this phase.

In the Elaboration/Application of Knowledge phase, teachers present students with activities that engage in higher-order thinking to create depth and breadth of student knowledge, while connecting ideas across topics within and across STEM. Students apply what they have learned thus far in the module to a new context or elaborate on what they have learned about the topic to a deeper level of detail.

In the last phase, Evaluation/Assessment, teachers give students summative feedback on their knowledge and skills as demonstrated through the challenge. This is not the only point of assessment (as discussed in the section on Embedded Formative Assessments), but it is an assessment of the culmination of the knowledge and skills for the module. Students demonstrate their cognitive growth at this point and reflect on how far they have come since the beginning of the module. The challenges are designed to be multidimensional in the ways students must collaborate and communicate their new knowledge.

STEM RESEARCH NOTEBOOK

One of the main components of the *STEM Road Map Curriculum Series* is the STEM Research Notebook, a place for students to capture their ideas, questions, observations, reflections, evidence of progress, and other items associated with their daily work. At the beginning of each module, the teacher walks students through the setup of the STEM

Research Notebook, which could be a three-ring binder, composition book, or spiral notebook. You may wish to have students create divided sections so that they can easily access work from various disciplines during the module. Electronic notebooks kept on student devices are also acceptable and encouraged. Students will develop their own table of contents and create chapters in the notebook for each module.

Each lesson in the *STEM Road Map Curriculum Series* includes one or more prompts that are designed for inclusion in the STEM Research Notebook and appear as questions or statements that the teacher assigns to students. These prompts require students to apply what they have learned across the lesson to solve the big problem or challenge for that module. Each lesson is designed to meaningfully refer students to the larger problem or challenge they have been assigned to solve with their teams. The STEM Research Notebook is designed to be a key formative assessment tool, as students' daily entries provide evidence of what they are learning. The notebook can be used as a mechanism for dialogue between the teacher and students, as well as for peer and self-evaluation.

The use of the STEM Research Notebook is designed to scaffold student notebooking skills across the grade bands in the *STEM Road Map Curriculum Series*. In the early grades, children learn how to organize their daily work in the notebook as a way to collect their products for future reference. In elementary school, students structure their notebooks to integrate background research along with their daily work and lesson prompts. In the upper grades (middle and high school), students expand their use of research and data gathering through team discussions to more closely mirror the work of STEM experts in the real world.

THE ROLE OF ASSESSMENT IN THE *STEM ROAD MAP CURRICULUM SERIES*

Starting in the middle years and continuing into secondary education, the word *assessment* typically brings grades to mind. These grades may take the form of a letter or a percentage, but they typically are used as a representation of a student's content mastery. If well thought out and implemented, however, classroom assessment can offer teachers, parents, and students valuable information about student learning and misconceptions that does not necessarily come in the form of a grade (Popham 2013).

The *STEM Road Map Curriculum Series* provides a set of assessments for each module. Teachers are encouraged to use assessment information for more than just assigning grades to students. Instead, assessments of activities requiring students to actively engage in their learning, such as student journaling in STEM Research Notebooks, collaborative presentations, and constructing graphic organizers, should be used to move student learning forward. Whereas other curriculum with assessments may include objective-type (multiple-choice or matching) tests, quizzes, or worksheets, we have intentionally avoided these forms of assessments to better align assessment strategies with teacher instruction and

student learning techniques. Since the focus of this book is on project- or problem-based STEM curriculum and instruction that focuses on higher-level thinking skills, appropriate and authentic performance assessments were developed to elicit the most reliable and valid indication of growth in student abilities (Brookhart and Nitko 2008).

Comprehensive Assessment System

Assessment throughout all STEM Road Map curriculum modules acts as a comprehensive system in which formative and summative assessments work together to provide teachers with high-quality information on student learning. Formative assessment occurs when the teacher finds out formally or informally what a student knows about a smaller, defined concept or skill and provides timely feedback to the student about his or her level of proficiency. Summative assessments occur when students have performed all activities in the module and are given a cumulative performance evaluation in which they demonstrate their growth in learning.

A comprehensive assessment system can be thought of as akin to a sporting event. Formative assessments are the practices: It is important to accomplish them consistently, they provide feedback to help students improve their learning, and making mistakes can be worthwhile if students are given an opportunity to learn from them. Summative assessments are the competitions: Students need to be prepared to perform at the best of their ability. Without multiple opportunities to practice skills along the way through formative assessments, students will not have the best chance of demonstrating growth in abilities through summative assessments (Black and Wiliam 1998).

Embedded Formative Assessments

Formative assessments in this module serve two main purposes: to provide feedback to students about their learning and to provide important information for the teacher to inform immediate instructional needs. Providing feedback to students is particularly important when conducting problem- or project-based learning because students take on much of the responsibility for learning, and teachers must facilitate student learning in an informed way. For example, if students are required to conduct research for the Activity/Exploration phase but are not familiar with what constitutes a reliable resource, they may develop misconceptions based on poor information. When a teacher monitors this learning through formative assessments and provides specific feedback related to the instructional goals, students are less likely to develop incomplete or incorrect conceptions in their independent investigations. By using formative assessment to detect problems in student learning and then acting on this information, teachers help move student learning forward through these teachable moments.

Formative assessments come in a variety of formats. They can be informal, such as asking students probing questions related to student knowledge or tasks or simply observing students engaged in an activity to gather information about student skills. Formative assessments can also be formal, such as a written quiz or a laboratory practical. Regardless of the type, three key steps must be completed when using formative assessments (Sondergeld, Bell, and Leusner 2010). First, the assessment is delivered to students so that teachers can collect data. Next, teachers analyze the data (student responses) to determine student strengths and areas that need additional support. Finally, teachers use the results from information collected to modify lessons and create learning environments that reinforce weak points in student learning. If student learning information is not used to modify instruction, the assessment cannot be considered formative in nature.

Formative assessments can be about content, science process skills, or even learning skills. When a formative assessment focuses on content, it assesses student knowledge about the disciplinary core ideas from the *Next Generation Science Standards* (*NGSS*) or content objectives from *Common Core State Standards for Mathematics* (*CCSS Mathematics*) or *Common Core State Standards for English Language Arts* (*CCSS ELA*). Content-focused formative assessments ask students questions about declarative knowledge regarding the concepts they have been learning. Process skills formative assessments examine the extent to which a student can perform science and engineering practices from the *NGSS* or process objectives from *CCSS Mathematics* or *CCSS ELA*, such as constructing an argument. Learning skills can also be assessed formatively by asking students to reflect on the ways they learn best during a module and identify ways they could have learned more.

Assessment Maps

Assessment maps or blueprints can be used to ensure alignment between classroom instruction and assessment. If what students are learning in the classroom is not the same as the content on which they are assessed, the resultant judgment made on student learning will be invalid (Brookhart and Nitko 2008). Therefore, the issue of instruction and assessment alignment is critical. The assessment map for this book (found in Chapter 3) indicates by lesson whether the assessment should be completed as a group or on an individual basis, identifies the assessment as formative or summative in nature, and aligns the assessment with its corresponding learning objectives.

Note that the module includes far more formative assessments than summative assessments. This is done intentionally to provide students with multiple opportunities to practice their learning of new skills before completing a summative assessment. Note also that formative assessments are used to collect information on only one or two learning objectives at a time so that potential relearning or instructional modifications can focus on smaller and more manageable chunks of information. Conversely, summative assessments in the module cover many more learning objectives, as they are traditionally used as final markers of student learning. This is not to say that information collected from summative assessments cannot or should not be used formatively. If teachers find that gaps in student learning persist after a summative assessment is completed, it is important to revisit these existing misconceptions or areas of weakness before moving on (Black et al. 2003).

SELF-REGULATED LEARNING THEORY IN THE STEM ROAD MAP MODULES

Many learning theories are compatible with the STEM Road Map modules, such as constructivism, situated cognition, and meaningful learning. However, we feel that the self-regulated learning theory (SRL) aligns most appropriately (Zimmerman 2000). SRL requires students to understand that thinking needs to be motivated and managed (Ritchhart, Church, and Morrison 2011). The STEM Road Map modules are student centered and are designed to provide students with choices, concrete hands-on experiences, and opportunities to see and make connections, especially across subjects (Eliason and Jenkins 2012; NAEYC 2016). Additionally, SRL is compatible with the modules because it fosters a learning environment that supports students' motivation, enables students to become aware of their own learning strategies, and requires reflection on learning while experiencing the module (Peters and Kitsantas 2010).

The theory behind SRL (see Figure 2.2) explains the different processes that students engage in before, during, and after a learning task. Because SRL is a cyclical learning process, the accomplishment of one cycle develops strategies for the next learning cycle. This cyclic way of learning aligns with the various sections in the STEM Road Map lesson plans on Introductory Activity/Engagement, Activity/Exploration, Explanation, Elaboration/Application of Knowledge, and Evaluation/Assessment. Since the students engaged in a module take on much of the responsibility for learning, this theory also provides guidance for teachers to keep students on the right track.

The remainder of this section explains how SRL theory is embedded within the five sections of each module and points out ways to

Figure 2.2. SRL Theory

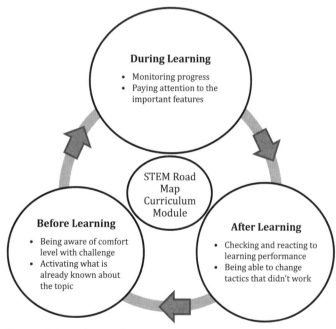

Source: Adapted from Zimmerman 2000.

support students in becoming independent learners of STEM while productively functioning in collaborative teams.

Before Learning: Setting the Stage

Before attempting a learning task such as the STEM Road Map modules, teachers should develop an understanding of their students' level of comfort with the process of accomplishing the learning and determine what they already know about the topic. When students are comfortable with attempting a learning task, they tend to take more risks in learning and as a result achieve deeper learning (Bandura 1986).

The STEM Road Map curriculum modules are designed to foster excitement from the very beginning. Each module has an Introductory Activity/Engagement section that introduces the overall topic from a unique and exciting perspective, engaging the students to learn more so that they can accomplish the challenge. The Introductory Activity also has a design component that helps teachers assess what students already know about the topic of the module. In addition to the deliberate designs in the lesson plans to support SRL, teachers can support a high level of student comfort with the learning challenge by finding out if students have ever accomplished the same kind of task and, if so, asking them to share what worked well for them.

During Learning: Staying the Course

Some students fear inquiry learning because they aren't sure what to do to be successful (Peters 2010). However, the STEM Road Map curriculum modules are embedded with tools to help students pay attention to knowledge and skills that are important for the learning task and to check student understanding along the way. One of the most important processes for learning is the ability for learners to monitor their own progress while performing a learning task (Peters 2012). The modules allow students to monitor their progress with tools such as the STEM Research Notebooks, in which they record what they know and can check whether they have acquired a complete set of knowledge and skills. The STEM Road Map modules support inquiry strategies that include previewing, questioning, predicting, clarifying, observing, discussing, and journaling (Morrison and Milner 2014). Through the use of technology throughout the modules, inquiry is supported by providing students access to resources and data while enabling them to process information, report the findings, collaborate, and develop 21st century skills.

It is important for teachers to encourage students to have an open mind about alternative solutions and procedures (Milner and Sondergeld 2015) when working through the STEM Road Map curriculum modules. Novice learners can have difficulty knowing what to pay attention to and tend to treat each possible avenue for information as equal (Benner 1984). Teachers are the mentors in a classroom and can point out ways for students to approach learning during the Activity/Exploration, Explanation, and

Elaboration/Application of Knowledge portions of the lesson plans to ensure that students pay attention to the important concepts and skills throughout the module. For example, if a student is to demonstrate conceptual awareness of motion when working on roller coaster research, but the student has misconceptions about motion, the teacher can step in and redirect student learning.

After Learning: Knowing What Works

The classroom is a busy place, and it may often seem that there is no time for self-reflection on learning. Although skipping this reflective process may save time in the short term, it reduces the ability to take into account things that worked well and things that didn't so that teaching the module may be improved next time. In the long run, SRL skills are critical for students to become independent learners who can adapt to new situations. By investing the time it takes to teach students SRL skills, teachers can save time later, because students will be able to apply methods and approaches for learning that they have found effective to new situations. In the Evaluation/Assessment portion of the STEM Road Map curriculum modules, as well as in the formative assessments throughout the modules, two processes in the after-learning phase are supported: evaluating one's own performance and accounting for ways to adapt tactics that didn't work well. Students have many opportunities to self-assess in formative assessments, both in groups and individually, using the rubrics provided in the modules.

The designs of the *NGSS* and *CCSS* allow for students to learn in diverse ways, and the STEM Road Map curriculum modules emphasize that students can use a variety of tactics to complete the learning process. For example, students can use STEM Research Notebooks to record what they have learned during the various research activities. Notebook entries might include putting objectives in students' own words, compiling their prior learning on the topic, documenting new learning, providing proof of what they learned, and reflecting on what they felt successful doing and what they felt they still needed to work on. Perhaps students didn't realize that they were supposed to connect what they already knew with what they learned. They could record this and would be prepared in the next learning task to begin connecting prior learning with new learning.

SAFETY IN STEM

Student safety is a primary consideration in all subjects but is an area of particular concern in science, where students may interact with unfamiliar tools and materials that may pose additional safety risks. It is important to implement safety practices within the context of STEM investigations, whether in a classroom laboratory or in the field. When you keep safety in mind as a teacher, you avoid many potential issues with the lesson while also protecting your students.

STEM safety practices encompass things considered in the typical science classroom. Ensure that students are familiar with basic safety considerations, such as wearing protective equipment (e.g., safety glasses or goggles and latex-free gloves) and taking care with sharp objects, and know emergency exit procedures. Teachers should learn beforehand the locations of the safety eyewash, fume hood, fire extinguishers, and emergency shut-off switch in the classroom and how to use them. Also be aware of any school or district safety policies that are in place and apply those that align with the work being conducted in the lesson. It is important to review all safety procedures annually.

STEM investigations should always be supervised. Each lesson in the modules includes teacher guidelines for applicable safety procedures that should be followed. Before each investigation, teachers should go over these safety procedures with the student teams. Some STEM focus areas such as engineering require that students can demonstrate how to properly use equipment in the maker space before the teacher allows them to proceed with the lesson.

Information about classroom science safety, including a safety checklist for science classrooms, general lab safety recommendations, and links to other science safety resources, is available at the Council of State Science Supervisors (CSSS) website at *www.csss-science. org/safety.shtml*. The National Science Teachers Association (NSTA) provides a list of science rules and regulations, including standard operating procedures for lab safety, and a safety acknowledgment form for students and parents or guardians to sign. You can access these resources at *http://static.nsta.org/pdfs/SafetyInTheScienceClassroom.pdf*. In addition, NSTA's Safety in the Science Classroom web page (*www.nsta.org/safety*) has numerous links to safety resources, including papers written by the NSTA Safety Advisory Board.

Disclaimer: The safety precautions for each activity are based on use of the recommended materials and instructions, legal safety standards, and better professional practices. Using alternative materials or procedures for these activities may jeopardize the level of safety and therefore is at the user's own risk.

REFERENCES

Bandura, A. 1986. *Social foundations of thought and action: A social cognitive theory*. Englewood Cliffs, NJ: Prentice-Hall.

Barell, J. 2006. *Problem-based learning: An inquiry approach*. Thousand Oaks, CA: Corwin Press.

Benner, P. 1984. *From novice to expert: Excellence and power in clinical nursing practice*. Menlo Park, CA: Addison-Wesley Publishing Company.

Black, P., C. Harrison, C. Lee, B. Marshall, and D. Wiliam. 2003. *Assessment for learning: Putting it into practice*. Berkshire, UK: Open University Press.

Black, P., and D. Wiliam. 1998. Inside the black box: Raising standards through classroom assessment. *Phi Delta Kappan* 80 (2): 139–148.

Blumenfeld, P., E. Soloway, R. Marx, J. Krajcik, M. Guzdial, and A. Palincsar. 1991. Motivating project-based learning: Sustaining the doing, supporting learning. *Educational Psychologist* 26 (3): 369–398.

Brookhart, S. M., and A. J. Nitko. 2008. *Assessment and grading in classrooms.* Upper Saddle River, NJ: Pearson.

Bybee, R., J. Taylor, A. Gardner, P. Van Scotter, J. Carlson, A. Westbrook, and N. Landes. 2006. *The BSCS 5E instructional model: Origins and effectiveness. http://science.education.nih.gov/houseofreps. nsf/b82d55fa138783c2852572c9004f5566/$FILE/Appendix?D.pdf.*

Eliason, C. F., and L. T. Jenkins. 2012. *A practical guide to early childhood curriculum.* 9th ed. New York: Merrill.

Johnson, C. 2003. Bioterrorism is real-world science: Inquiry-based simulation mirrors real life. *Science Scope* 27 (3): 19–23.

Krajcik, J., and P. Blumenfeld. 2006. Project-based learning. In *The Cambridge handbook of the learning sciences,* ed. R. Keith Sawyer, 317–334. New York: Cambridge University Press.

Lambros, A. 2004. *Problem-based learning in middle and high school classrooms: A teacher's guide to implementation.* Thousand Oaks, CA: Corwin Press.

Milner, A. R., and T. Sondergeld. 2015. Gifted urban middle school students: The inquiry continuum and the nature of science. *National Journal of Urban Education and Practice* 8 (3): 442–461.

Morrison, V., and A. R. Milner. 2014. Literacy in support of science: A closer look at cross-curricular instructional practice. *Michigan Reading Journal* 46 (2): 42–56.

National Association for the Education of Young Children (NAEYC). 2016. Developmentally appropriate practice position statements. *www.naeyc.org/positionstatements/dap.*

Peters, E. E. 2010. Shifting to a student-centered science classroom: An exploration of teacher and student changes in perceptions and practices. *Journal of Science Teacher Education* 21 (3): 329–349.

Peters, E. E. 2012. Developing content knowledge in students through explicit teaching of the nature of science: Influences of goal setting and self-monitoring. *Science and Education* 21 (6): 881–898.

Peters, E. E., and A. Kitsantas. 2010. The effect of nature of science metacognitive prompts on science students' content and nature of science knowledge, metacognition, and self-regulatory efficacy. *School Science and Mathematics* 110: 382–396.

Popham, W. J. 2013. *Classroom assessment: What teachers need to know.* 7th ed. Upper Saddle River, NJ: Pearson.

Ritchhart, R., M. Church, and K. Morrison. 2011. *Making thinking visible: How to promote engagement, understanding, and independence for all learners.* San Francisco, CA: Jossey-Bass.

Sondergeld, T. A., C. A. Bell, and D. M. Leusner. 2010. Understanding how teachers engage in formative assessment. *Teaching and Learning* 24 (2): 72–86.

Zimmerman, B. J. 2000. Attaining self-regulation: A social-cognitive perspective. In *Handbook of self-regulation,* ed. M. Boekaerts, P. Pintrich, and M. Zeidner, 13–39. San Diego: Academic Press.

PART 2

SWING SET MAKEOVER

STEM ROAD MAP MODULE

SWING SET MAKEOVER MODULE OVERVIEW

Paula Schoeff, Janet B. Walton, Carla C. Johnson, and Erin Peters-Burton

THEME: The Represented World

LEAD DISCIPLINE: Science

MODULE SUMMARY

In this module, students examine the STEM aspects involved in constructing a new and improved design for playground equipment, referred to as *swing sets* throughout the module. Science inquiry activities allow students to explore the impact of balanced and unbalanced forces on motion. Students gain a conceptual understanding of motion and force when they relate how the body moves to the objects they have explored, paying close attention to the forces each exerts and the positions maintained to keep balance. In mathematics, students identify geometric shapes, collect data using mathematical tools, and then record and analyze data using line plots and bar graphs to explain and make predictions. Students are challenged to apply this knowledge in the Swing Set Makeover Design Challenge. In this challenge, students examine a variety of swing sets and respond to a survey. They work collaboratively using the engineering design process (EDP) to sketch and develop a small-scale model of their proposed design, using geometric shapes and precise measurements. Finally, students draft a blog or newsletter detailing the key components of their design and justifying how their design is an improvement on existing swing set designs (adapted from Capobianco et al. 2015).

ESTABLISHED GOALS AND OBJECTIVES

At the conclusion of this module, students will be able to do the following:

- Use the EDP to design a swing set

- Compare and contrast swing set designs as they improve on or diminish a "fun factor" score on a survey

- Recognize and describe gravity and friction as forces

- Evaluate forces that interact with the body and swing set unit on the playground and associate these forces with design challenges

- Evaluate and recommend alternative shapes and designs for the swing set

- Design and implement an investigation around a testable question

- Use data analysis as evidence to answer a question under investigation

- Evaluate and implement safety features and materials of a swing set design

- Write a blog that describes the swing set design and the strengths of the design

- Recognize that citizens have civic responsibilities, which include helping care for public parks

- List ways that we, as citizens, can help care for our public parks

CHALLENGE OR PROBLEM FOR STUDENTS TO SOLVE: SWING SET MAKEOVER DESIGN CHALLENGE

Student teams are challenged to survey existing playground equipment (referred to as *swing sets* throughout the module), compare and contrast different designs in light of safety concerns connected to the playground equipment and its environment, and then sketch and build a small-scale model of their proposed design using geometric shapes and precise measurements. Details of the swing set and evidence of its improvement on existing playground equipment are synthesized in a blog or newsletter.

Students create swing set designs that will

- include a swing, a slide, and some sort of connecting partition;

- take advantage of features that maximize lessons about force and motion; and

- use fun features outlined in discussions while maintaining high standards of safety.

Driving Question: How can I use what I know about force and motion to create a plan and build a model of a swing set that is both fun and safe?

CONTENT STANDARDS ADDRESSED IN THIS STEM ROAD MAP MODULE

A full listing with descriptions of the standards this module addresses can be found in the appendix. Listings of the particular standards addressed within lessons are provided in a table for each lesson in Chapter 4.

STEM RESEARCH NOTEBOOK

Each student should maintain a STEM Research Notebook, which will serve as a place for students to organize their work throughout this module (see p. 12 for more general discussion on setup and use of the notebook). All written work in the module should be included in the notebook, including records of students' thoughts and ideas, fictional accounts based on the concepts in the module, and records of student progress through the EDP. The notebooks may be maintained across subject areas, giving students the opportunity to see that although their classes may be separated during the school day, the knowledge they gain is connected. You may also wish to have students include the STEM Research Notebook Guidelines student handout on page 26 in their notebooks.

Emphasize to students the importance of organizing all information in a Research Notebook. Explain to them that scientists and other researchers maintain detailed Research Notebooks in their work. These notebooks, which are crucial to researchers' work because they contain critical information and track the researchers' progress, are often considered legal documents for scientists who are pursuing patents or wish to provide proof of their discovery process.

STUDENT HANDOUT

STEM RESEARCH NOTEBOOK GUIDELINES

STEM professionals record their ideas, inventions, experiments, questions, observations, and other work details in notebooks so that they can use these notebooks to help them think about their projects and the problems they are trying to solve. You will each keep a STEM Research Notebook during this module that is like the notebooks that STEM professionals use. In this notebook, you will include all your work and notes about ideas you have. The notebook will help you connect your daily work with the big problem or challenge you are working to solve.

It is important that you organize your notebook entries under the following headings:

1. **Chapter Topic or Title of Problem or Challenge:** You will start a new chapter in your STEM Research Notebook for each new module. This heading is the topic or title of the big problem or challenge that your team is working to solve in this module.

2. **Date and Topic of Lesson Activity for the Day:** Each day, you will begin your daily entry by writing the date and the day's lesson topic at the top of a new page. Write the page number both on the page and in the table of contents.

3. **Information Gathered From Research:** This is information you find from outside resources such as websites or books.

4. **Information Gained From Class or Discussions With Team Members:** This information includes any notes you take in class and notes about things your team discusses. You can include drawings of your ideas here, too.

5. **New Data Collected From Investigations:** This includes data gathered from experiments, investigations, and activities in class.

6. **Documents:** These are handouts and other resources you may receive in class that will help you solve your big problem or challenge. Paste or staple these documents in your STEM Research Notebook for safekeeping and easy access later.

7. **Personal Reflections:** Here, you record your own thoughts and ideas on what you are learning.

8. **Lesson Prompts:** These are questions or statements that your teacher assigns you within each lesson to help you solve your big problem or challenge. You will respond to the prompts in your notebook.

9. **Other Items:** This section includes any other items your teacher gives you or other ideas or questions you may have.

MODULE LAUNCH

Students will engage in a group discussion about the swing sets at school or their neighborhood park to activate prior knowledge, share personal experiences, and highlight their ideas of what components they enjoy most. After the discussion, teams will be given a picture of a swing set that they will use to generate a list of criteria to analyze the swing set for fun factors on a scale of 1 to 5 (see the Fun Factor Survey on p. 77). After the teams complete their tables, the class will visit the school playground swing set and use the same scale to determine its score on the survey.

Introduce the challenge by telling students that after they explore force and motion, their challenge will be to act as mechanical engineers to create a model of a swing set that will meet their standards for fun and safety. After they have developed their models, they will share their designs and defend their reasoning for design decisions online in a blog.

PREREQUISITE SKILLS FOR THE MODULE

Students enter this module with a wide range of preexisting skills, information, and knowledge. Table 3.1 (p. 28) provides an overview of prerequisite skills and knowledge that students are expected to apply in this module, along with examples of how they apply this knowledge throughout the module. Differentiation strategies are also provided for students who may need additional support in acquiring or applying this knowledge.

Table 3.1. Prerequisite Key Knowledge and Examples of Applications and Differentiation Strategies

Prerequisite Key Knowledge	Application of Knowledge by Students	Differentiation for Students Needing Knowledge
Science Knowledge • Force is a push or pull. • Gravity pulls objects down (toward the Earth). • Friction causes objects to slow down. • Squares and rectangles have length and width measurements. All four sides are added to find the perimeter and the length and width are multiplied to find the area. • Scientists and engineers solve problems in a similar fashion. • Some spaces are public spaces.	*Science Knowledge* • Experience, model, and describe gravitational force on an inclined plane using a variety of angles and materials. • See the impact of friction when testing a variety of materials on the ramp. • Experience, model, and describe rotational force using a variety of types of pendulums. • Design a swing set that demonstrates careful examination of materials and how the design is affected by gravitational force.	*Science Knowledge* • Provide adequate scaffolding to allow students to make informed decisions when designing tests. • Provide opportunities to work in teams as students draft sketches of their models and tests. • Provide opportunities to use oral and written language to describe scientific processes and principles.
Inquiry Skills • Ask questions, make logical predictions, plan investigations, and represent data. • Use senses and tools to make observations. • Communicate and plan simple investigations. • Communicate understanding of data using age-appropriate vocabulary.	*Inquiry Skills* • Select and use appropriate tools and equipment to conduct an investigation. • Identify tools needed to investigate specific questions. • Maintain a notebook that includes observations, data, diagrams, and reflections. • Analyze and communicate findings from multiple investigations of similar phenomena to reach a conclusion.	*Inquiry Skills* • Model selection and use appropriate tools and simple equipment to conduct an investigation. • Provide samples of a STEM Research Notebook. • Scaffold student efforts to organize data into appropriate tables, graphs, drawings, or diagrams by providing step-by-step instructions. • Identify specific investigations that could be used to answer a particular question and identify reasons for this choice.

Continued

Table 3.1. (*continued*)

Prerequisite Key Knowledge	Application of Knowledge by Students	Differentiation for Students Needing Knowledge
Measurement Skills • Measuring distance using standard and metric units. • Measuring distance to the nearest inch, half inch, and one-quarter inch. • Evaluating time to the nearest minute and second.	*Measurement Skills* • Measure distances using a tape measure with metric units. • Calculate distances using methods developed through experimentation. • Use stopwatches and other strategies to determine periods of motion. • Calculate speeds in units they create as well as cm units over a period of time.	*Measurement Skills* • Provide students with opportunities to practice measuring distances using various units and measuring time. • Provide students with additional content, including textbook support, teacher instruction, and online videos for using stopwatches and explaining strategies for measuring time. • Provide instruction in use of a stopwatch and identifying the units as seconds and minutes.
Numbers and Representation • Represent whole numbers up to and including 1,000. • Add and subtract whole numbers. • Multiply and divide whole numbers. • Create real data and represent findings in appropriately labeled tables. • Represent data using line plots.	*Numbers and Representation* • Add, subtract, multiply, and divide numbers to analyze findings and make decisions. • Calculate distances and speeds and track the results in tables. • Support design decisions using numbers in tables and graphs (line plots).	*Numbers and Representation* • Review and provide models of addition and subtraction up to 1,000. • Review multiplication and division using examples of distance and time. • Use textbook support, teacher instruction, models, graphic organizers, and online videos to provide practice using tables and graphs.
Geometry • Recognize and identify geometric shapes and patterns. • Recognize that a plane is composed of geometric shapes and patterns. • Measure perimeter and area of geometric shapes.	*Geometry* • Find geometric shapes and patterns in the playground and playground swing sets. • Experiment with geometric shapes while designing playground equipment. • Consider the space that is available and the size of the apparatus being designed (perimeter and area).	*Geometry* • Review and provide tessellation models and manipulatives. • Put together puzzles composed of geometric shapes to get a sense of how shapes make up planes. • Practice measuring shapes around the classroom.

Continued

Table 3.1. (*continued*)

Prerequisite Key Knowledge	Application of Knowledge by Students	Differentiation for Students Needing Knowledge
Reading • Use information gained from illustrations and words in print or digital text to build understanding of scientific concepts. • Use information gained from illustrations and words in print or digital text to build understanding of safe and dangerous conditions on a playground.	*Reading* • Describe the relationship between a group of images and descriptions using language that describes where, when, why, and how as it pertains to playgrounds and safe versus unsafe conditions.	*Reading* • Provide reading strategies to support comprehension of nonfiction texts, including using vocabulary notecards, creating graphic organizers, writing in the STEM Research Notebook, and discussions.
Writing • Use science terms to inform and explain thoughts and ideas about the topic. • Use key terminology as words and pictures.	*Writing* • Write informative and explanatory narratives to convey ideas and information clearly. • Write narratives to describe experiences using effective techniques, descriptive details, and clear event sequences.	*Writing* • Provide a template for writing. • Provide writing organization handouts to scaffold student work. • Provide rubrics that have a consistent format so students can measure their own writing.
Communication Skills • Participate in collaborative conversations using appropriate language and skills. • Effectively support scientific knowledge with appropriate language and relevant, descriptive details.	*Communication Skills* • Engage in a number of collaborative discussions that convey and support learning. • Write a blog to describe the swing set design and support it with scientific reasoning.	*Communication Skills* • Scaffold student understanding of communication skills by providing examples of appropriate language and presentation. • Provide handouts and rubrics to support organization of facts and use of relevant descriptive details.

POTENTIAL STEM MISCONCEPTIONS

Students enter the classroom with a wide variety of prior knowledge and ideas, so it is important to be alert to misconceptions, or inappropriate understandings of foundational knowledge. These misconceptions can be classified as one of several types: "preconceived notions," opinions based on popular beliefs or understandings; "nonscientific beliefs," knowledge students have gained about science from sources outside the scientific community; "conceptual misunderstandings," incorrect conceptual models based on incomplete understanding of concepts; "vernacular misconceptions," misunderstandings of words based on their common use versus their scientific use; and "factual misconceptions," incorrect or imprecise knowledge learned in early life that remains unchallenged (NRC 1997, p. 28). Misconceptions must be addressed and dismantled for students to reconstruct their knowledge, and therefore teachers should be prepared to take the following steps:

- *Identify students' misconceptions.*

- *Provide a forum for students to confront their misconceptions.*

- *Help students reconstruct and internalize their knowledge, based on scientific models.
  (NRC 1997, p. 29)*

Keeley and Harrington (2010) recommend using diagnostic tools such as probes and formative assessment to identify and confront student misconceptions and begin the process of reconstructing student knowledge. Keeley's *Uncovering Student Ideas in Science* series contains probes targeted toward uncovering student misconceptions in a variety of areas and may be useful resources for addressing student misconceptions in this module.

Some commonly held misconceptions specific to lesson content are provided with each lesson so that you can be alert for student misunderstanding of the science concepts presented and used during this module. The American Association for the Advancement of Science has also identified misconceptions that students frequently hold regarding various science concepts (see the links at *http://assessment.aaas.org/topics*).

SRL PROCESS COMPONENTS

Table 3.2 illustrates some activities in the Swing Set Makeover module and how they align with the self-regulated learning (SRL) process before, during, and after learning.

Table 3.2. SRL Process Components

Learning Process Components	Example From Swing Set Makeover Module	Lesson Number and Learning Component
BEFORE LEARNING		
Motivates students	Students discuss their ideas with the class of the most enjoyable components of the swing sets they have experienced.	Lesson 1, Introductory Activity/Engagement
Evokes prior learning	Students engage in a group discussion about the swing sets at school or their neighborhood and share their personal experiences. Teacher relates this information to the challenge.	Lesson 1, Activity/ Exploration
DURING LEARNING		
Focuses on important features	Students revisit the Fun Factor Survey from the launch and draw a sketch of a slide for a swing set, using their skills and knowledge learned about Newton's first law of motion and geometric figures.	Lesson 2, Activity/ Exploration
Helps students monitor their progress	Students share their sketches and proposals for their slide designs.	Lesson 2, Activity/ Exploration
AFTER LEARNING		
Evaluates learning	After student teams plan the arrangement of the swing set on the site, identify the parts, choose the materials, and build the model, they test their model and document the process.	Lesson 4, Activity/ Exploration
Takes account of what worked and what did not work	Students reflect on what works with their original design and refine the model. Then they defend their design decisions by communicating them in a blog.	Lesson 4, Activity/ Exploration

STRATEGIES FOR DIFFERENTIATING INSTRUCTION WITHIN THIS MODULE

For the purposes of this curriculum module, differentiated instruction is conceptualized as a way to tailor instruction—including process, content, and product—to various student needs in your class. A number of differentiation strategies are integrated into lessons across the module. The problem- and project-based learning approach used in the lessons are designed to address students' multiple intelligences by providing a variety of entry points and methods to investigate the key concepts in the module (for example, investigating playground equipment from the perspectives of science, safety, and fun via scientific inquiry, literature, journaling, and collaborative design). Differentiation strategies for students needing support in prerequisite knowledge can be found in Table 3.1 (p. 28). You are encouraged to use information gained about student prior knowledge during introductory activities and discussions to inform your instructional differentiation. Strategies incorporated into this lesson include flexible grouping, varied environmental learning contexts, assessments, compacting, tiered assignments and scaffolding, and mentoring.

Flexible Grouping. Students work collaboratively in a variety of activities throughout this module. Grouping strategies you may choose to employ include student-led grouping, placing students in groups according to ability level, grouping students randomly, grouping them so that students in each group have complementary strengths (for instance, one student might be strong in mathematics, another in art, and another in writing), or grouping students according to common interests.

Varied Environmental Learning Contexts. Students have the opportunity to learn in various contexts throughout the module, including alone, in groups, in quiet reading and research-oriented activities, and in active learning in inquiry and design activities. In addition, students learn in a variety of ways through doing inquiry activities, journaling, reading a variety of texts, watching videos, class discussion, and conducting web-based research.

Assessments. Students are assessed in a variety of ways throughout the module, including individual and collaborative formative and summative assessments. Students have the opportunity to produce work via written text, oral and media presentations, and modeling. You may choose to provide students with additional choices of media for their products (for example, electronic slide presentations, posters, or student-created websites or blogs).

Compacting. Based on student prior knowledge, you may wish to adjust instructional activities for students who exhibit prior mastery of a learning objective. Because student work in science is largely collaborative throughout the module, this strategy may be most appropriate for mathematics, English language arts (ELA), or social studies activities. You may wish to compile a classroom database of research resources and supplementary

readings for a variety of reading levels and on a variety of topics related to the module's topic to provide opportunities for students to undertake independent reading.

Tiered Assignments and Scaffolding. Based on your awareness of student ability, understanding of concepts, and mastery of skills, you may wish to provide students with variations on activities by adding complexity to assignments or providing more or fewer learning supports for activities throughout the module. For instance, some students may need additional support in identifying key search words and phrases for web-based research or may benefit from cloze sentence handouts to enhance vocabulary understanding. Other students may benefit from expanded reading selections and additional reflective writing or from working with manipulatives and other visual representations of mathematical concepts. You may also work with your school librarian to compile a set of topical resources at a variety of reading levels.

Mentoring. As group design teamwork becomes increasingly complex throughout the module, you may wish to have a resource teacher, older student, or parent volunteer work with groups that struggle to stay on task and collaborate effectively.

STRATEGIES FOR ENGLISH LANGUAGE LEARNERS

Students who are developing proficiency in English language skills require additional supports to simultaneously learn academic content and the specialized language associated with specific content areas. WIDA (2012) has created a framework for providing support to these students and makes available rubrics and guidance on differentiating instructional materials for English language learners (ELLs). In particular, ELL students may benefit from additional sensory supports such as images, physical modeling, and graphic representations of module content, as well as interactive support through collaborative work. This module incorporates a variety of sensory supports and offers ongoing opportunities for ELL students to work collaboratively. The focus on playground equipment affords an opportunity for ELL students to share culturally diverse experiences with parks and playgrounds.

When differentiating instruction for ELL students, you should carefully consider the needs of these students as you introduce and use academic language in various language domains (listening, speaking, reading, and writing) throughout this module. To adequately differentiate instruction for ELL students, you should have an understanding of the proficiency level of each student. The following five overarching preK–5 WIDA learning standards are relevant to this module:

- Standard 1: Social and Instructional Language. Focus on following directions, personal information, and collaboration with peers.

- Standard 2: The language of Language Arts. Focus on nonfiction, fiction, sequence of story, and elements of story.

- Standard 3: The language of Mathematics. Focus on basic operations, number sense, interpretation of data, and patterns.

- Standard 4: The language of Science. Focus on forces in nature, scientific process, Earth and sky, living and nonliving things, organisms and environment, and weather.

- Standard 5: The language of Social Studies. Focus on community workers, homes and habitats, jobs and careers, and representations of Earth (maps and globes).

SAFETY CONSIDERATIONS FOR THE ACTIVITIES IN THIS MODULE

Student safety is a primary consideration in all subjects where students may interact with tools and materials with which they are unfamiliar and which may pose additional safety risks. You should ensure that your classroom set-up is in accord with your school's safety policies and that students are familiar with basic safety procedures, the location of protective equipment (e.g., safety glasses, gloves), and emergency exit procedures. For more general safety guidelines, see the Safety in STEM section in Chapter 2 (p. 18).

Internet safety is also important. You should develop an internet/blog protocol with students if guidelines are not already in place. Since students will use the internet for their research to acquire the needed data, you should monitor students' access to ensure that they are accessing only websites that you have clearly identified. Further, you should inform parents or guardians that students will create blogs and that you will closely monitor these projects. It is recommended that you not allow any website posts created by students to go public without first approving them.

During this module, students will be spending time outside exploring your school's playground. Before conducting these outdoor explorations, instruct students in safe practices for exploring the playground area.

DESIRED OUTCOMES AND MONITORING SUCCESS

The desired outcome for this module is outlined in Table 3.3, along with suggested ways to gather evidence to monitor student success. For more specific details on desired outcomes, see the Established Goals and Objectives sections for the module and individual lessons.

Table 3.3. Desired Outcome and Evidence of Success in Achieving Identified Outcome

Desired Outcome	Evidence of Success	
	Performance Tasks	Other Measures
Students apply an understanding of balanced and unbalanced forces and predict the motion of unbalanced forces that are applied to an object.	• Students maintain STEM Research Notebooks that contain graphic organizers with lab test data, sketches, research notes, and ELA-related work. • Students design a model of a swing set. • Students defend their design decisions in a blog or newsletter by making correlations to their observations from the unit. • Students are assessed using project rubrics that focus on learning and application of skills related to the academic content.	• STEM Research Notebooks are assessed using a STEM Research Notebook rubric. Clarify expectations and when necessary reproduce rubrics. • Student collaboration is evaluated using a self-assessment reflection form and peer feedback.

ASSESSMENT PLAN OVERVIEW AND MAP

Table 3.4 provides an overview of the major group and individual *products* and *deliverables,* or things that comprise the assessment for this module. See Table 3.5 (p. 38) for a full assessment map of formative and summative assessments in this module.

Table 3.4. Major Products and Deliverables in Lead Disciplines for Groups and Individuals

Lesson	Major Group Products and Deliverables	Major Individual Products and Deliverables
1	• Fun Factor Survey • Forces Push Back handout	• Bar graph • STEM Research Notebook entries
2	• Slide Makeover Sketch • Park Exhibition Showcase (research, collection of artifacts, and presentation)	• Ramp Investigation handouts • Slide Makeover Plan Book • Opinion Essay • STEM Research Notebook entries
3	• Swing Makeover Sketch • Jigsaw Research Template	• Pendulum Investigation handouts • Swing Makeover Plan Book • STEM Research Notebook entries
4	• Swing Set Makeover Model • Proposal Presentation • Blog or newsletter that provides evidence to support decisions for final swing set design	• STEM Research Notebook entries • Blog or newsletter contributions

Table 3.5. Assessment Map for Swing Set Makeover Module

Lesson	Assessment	Group/ Individual	Formative/ Summative	Lesson Objective Assessed
1	STEM Research Notebook *prompt*	Individual	Formative	• Recognize and describe gravity as a force. • Recognize and describe friction as a force.
1	Fun Factor Survey *handout*	Group	Formative	• Evaluate the impact of a swing set on its "fun factor" score.
1	Forces Push Back *handout*	Group	Formative	• Recognize and describe gravity as a force. • Recognize and describe friction as a force. • Evaluate a variety of activities to identify the forces that affect motion.
1	Bar Graph *rubric*	Individual/ Group	Summative	• Create bar graph with appropriate features.
1	Park Presentation *rubric*	Group	Formative	• Research a local park.
2	Ramp Investigations *handouts*	Individual	Formative	• Design and implement an investigation around a testable question. • Use data analysis as evidence to answer a question under investigation. • Relate the angle and length of a ramp to increased/ decreased speed.
2	Slide Makeover Plan Book *handout*	Individual	Formative	• Compare and contrast materials and angles to make decisions for a new slide design.
2	Slide Design Sketch *rubric*	Group	Formative	• Explain that friction causes work (energy) to be wasted when objects go down the slide. • Compare and contrast materials and angles to make decisions for a new slide design. • Critique slide designs of classmates to provide constructive feedback.

Continued

Table 3.5. (*continued*)

Lesson	Assessment	Group/ Individual	Formative/ Summative	Lesson Objective Assessed
2	Geometry Scavenger Hunt *handout*	Group	Formative	• Calculate the area of a surface. • Identify the geometric shapes that appear in a swing set.
2	Writing the OREO Way *handout*	Individual	Formative	• Formulate opinion statements and provide supporting reasons for opinions.
2	Park Presentation *rubric*	Group	Formative	• Present research on local park.
2	STEM Research Notebook *prompt*	Individual	Formative	• Understand area and perimeter.
3	Pendulum Investigations *handouts and rubric*	Individual	Formative	• Recognize and describe gravity and inertia's influence on pendulum motion. • Analyze the motion of a pendulum to make predictions.
3	Swing Makeover Plan Book *handout*	Individual	Formative	• Compare and contrast materials and angles to make decisions for a new slide design.
3	Jigsaw Research *template*	Individual	Formative	• Understand that components of swing design include motion, safety, materials, and aesthetics.
3	Swing Design Sketch *rubric*	Group	Formative	• Predict and evaluate the impact of design and materials on the swing's fun factor rating. • Understand that components of swing design include motion, safety, materials, and aesthetics. • Design a swing for the Swing Set Makeover Design Challenge.

Continued

Table 3.5. (*continued*)

Lesson	Assessment	Group/ Individual	Formative/ Summative	Lesson Objective Assessed
4	Model for Swing Set Makeover *handouts and rubrics*	Group	Summative	• Build a model for a new swing set that considers shapes and forces and has a high rating on the Fun Factor Survey from Lesson 1. • Build a model for a new swing set that incorporates an understanding of basic safety features. • Effectively use shapes, materials, and measurements that affect speed, aesthetics, and safety on a new swing set design.
4	Blog/ Newsletter *handouts and rubrics*	Group	Summative	• Create a blog or newsletter that addresses the design decisions made in the design process and the benefits of the swing set design.
4	Proposal Presentation *rubric*	Group	Summative	• Communicate information about swing set designs in a presentation.

MODULE TIMELINE

Tables 3.6–3.10 (pp. 41–43) provide lesson timelines for each week of the module. The timelines are provided for general guidance only and are based on class times of approximately 45 minutes.

Table 3.6. STEM Road Map Module Schedule for Week One

Day 1	Day 2	Day 3	Day 4	Day 5
Lesson 1 *Forces Push Back* • Launch the module and introduce the Fun Factor Survey as a tool to analyze swing sets. • In ELA, students begin to explore module concepts through fiction and nonfiction literature.	*Lesson 1* *Forces Push Back* • Students explore personal preferences for swing sets and work with their design teams to generate team standards and a team "fun factor" rubric.	*Lesson 1* *Forces Push Back* • Teams use their fun factor rubrics to evaluate playground images. • Graphing activities support understanding of fun factor data. • Launch a team social studies park research project to explore parks in the area and investigate social responsibilities to the community.	*Lesson 1* *Forces Push Back* • Explore forces associated with playground activities. • Identify how gravity and friction forces affect motion.	*Lesson 2* *Slippery Slide Design* • Introduce balanced and unbalanced forces. • Introduce the use of geometric shapes in buildings and playground equipment.

Table 3.7. STEM Road Map Module Schedule for Week Two

Day 6	Day 7	Day 8	Day 9	Day 10
Lesson 2 *Slippery Slide Design* • Investigate balanced and unbalanced forces using inclined planes. • Explore geometric shapes on the school playground, take measurements, and create maps of the playground.	*Lesson 2* *Slippery Slide Design* • Investigate balanced and unbalanced forces using inclined planes. • Continue to explore geometric shapes on the school playground, take measurements, and create maps of the playground.	*Lesson 2* *Slippery Slide Design* • Introduce annotated sketching. • Teams begin their Slide Makeover Plan Books.	*Lesson 2* *Slippery Slide Design* • Teams work on their Slide Makeover Plan Books. • In ELA, students begin to write opinion essays.	*Lesson 2* *Slippery Slide Design* • Teams complete their Slide Makeover Plan Books and provide feedback to other teams on their designs. • In social studies, students present their park research projects.

Table 3.8. STEM Road Map Module Schedule for Week Three

Day 11	Day 12	Day 13	Day 14	Day 15
Lesson 3	*Lesson 3*	*Lesson 3*	*Lesson 3*	*Lesson 3*
Swinging Pendulums	*Swinging Pendulums*	*Swinging Pendulums*	*Swinging Pendulums*	*Swinging Pendulums*
• Introduce pendulum motion and pendulum activities to explore and predict the behavior of swinging objects.	• Students participate in a pendulum activity. • In ELA, students begin a jigsaw research activity to investigate topics related to playgrounds.	• Introduce scaled drawings. • Teams begin their Swing Makeover Plan Books.	• Teams continue working on their Swing Makeover Plan Books.	• Teams complete their Swing Makeover Plan Books and provide feedback to other teams on their designs.

Table 3.9. STEM Road Map Module Schedule for Week Four

Day 16	Day 17	Day 18	Day 19	Day 20
Lesson 4	*Lesson 4*	*Lesson 4*	*Lesson 4*	*Lesson 4*
Swing Set Makeover Design Challenge	*Swing Set Makeover Design Challenge*	*Swing Set Makeover Design Challenge*	*Swing Set Makeover Design Challenge*	*Swing Set Makeover Design Challenge*
• Introduce the engineering design process (EDP). Students participate in an EDP activity. • In ELA, introduce blogging and the blog (or newsletter) students will create for their swing set models and design process.	• Teams create a layout of the site for their swing set models.	• Teams choose materials and begin to construct their swing set models.	• Teams continue to build their swing set models.	• Teams continue to build their swing set models.

Table 3.10. STEM Road Map Module Schedule for Week Five

Day 21	Day 22	Day 23	Day 24	Day 25
Lesson 4 *Swing Set Makeover Design Challenge* • Teams test and refine their swing set models.	*Lesson 4* *Swing Set Makeover Design Challenge* • Teams document their design process and details in blogs or newsletters.	*Lesson 4* *Swing Set Makeover Design Challenge* • Teams present their models and discuss and defend design decisions in presentations.	*Lesson 4* *Swing Set Makeover Design Challenge* • These days are left open to accommodate Lesson 4 activities that may have taken longer than anticipated. If the module is complete by day 23, options include a field trip to a local park or science museum or guest speakers.	

RESOURCES

The media specialist can help teachers locate resources for students to view and read about recreational equipment, parks, and related physics content. Special educators and reading specialists can help find supplemental sources for students needing extra support in reading and writing. Additional resources may be found online. Community resources for this module may include mechanical engineers, representatives of parks departments, and playground manufacturing representatives.

REFERENCES

Capobianco, B. M., C. Parker, A. Laurier, and J. Rankin. 2015. The STEM road map for grades 3–5. In *STEM Road Map: A framework for integrated STEM education,* ed. C. C. Johnson, E. E. Peters-Burton, and T. J. Moore, 68–95. New York: Routledge. *www.routledge.com/products/9781138804234.*

Keeley, P., and R. Harrington. 2010. *Uncovering student ideas in physical science, volume 1: 45 new force and motion assessment probes.* Arlington, VA: NSTA Press.

National Research Council (NRC). 1997. *Science teaching reconsidered: A handbook.* Washington, DC: National Academies Press.

WIDA. 2012. 2012 amplification of the English language development standards: Kindergarten–grade 12. *https://wida.wisc.edu/teach/standards/eld.*

SWING SET MAKEOVER LESSON PLANS

Paula Schoeff, Janet B. Walton, Carla C. Johnson, Erin Peters-Burton

Lesson Plan 1: Forces Push Back

This lesson introduces students to the module and the culminating challenge of the module, the Swing Set Makeover Design Challenge. A video, discussions, and a trip to the school playground help students connect to the project and excite their curiosity. Science activities focus on motion, emphasizing inertia, gravity, and friction forces. In mathematics, students graph data related to swing sets, and in English language arts (ELA), the class begins to explore parks through literature. Social studies conversations launch a park research project and address citizens' responsibilities to use their expertise and knowledge to benefit the community.

ESSENTIAL QUESTIONS

- What forces are at work on a playground?

- How do the design and shapes in a swing set affect speed and motion?

- What is the relationship of inertia and other forces in playground play activities?

- How does the force called friction slow or decrease gravity's pull?

- What are some ways that friction slows objects such as balls or Frisbees when nothing seems to be touching the object?

- How can what's learned about inertia, gravity, and friction be applied to a slide design?

ESTABLISHED GOALS AND OBJECTIVES

At the conclusion of this lesson, students will be able to do the following:

- Evaluate the impact of a swing set on its "fun factor" score

- Recognize and describe gravity as a force

- Recognize and describe friction as a force

- Evaluate a variety of activities to identify the forces that affect motion

- Create a bar graph with appropriate features

- Research a local park

TIME REQUIRED

- 4 days (approximately 45 minutes each day; see Table 3.6, p. 41)

MATERIALS

Handouts for Lesson 1

- Playground Swing Set images (1 per team)

- Fun Factor Survey handout (1 per student)

- Fun Factor Bar Graph Template—optional (1 per student)

- Forces Push Back handout (1 per team)

- Park Exhibition Showcase handout (1 per team)

Rubrics for Lesson 1

- STEM Research Notebook Rubric

- Bar Graph Rubric

- Park Presentation Rubric

Necessary Materials for Lesson 1

- STEM Research Notebooks (for each student)

- Internet access for showing swing set images and for student research

- Chart paper for a KWL (Know, Want to Know, Learned) chart

- Optional: Pictures of swing sets, slides, ramps, pendulums, or other images on display to create excitement and for students to use as inspiration when creating the cover pages for their STEM Research Notebooks

Additional Materials for Fun Factor Survey

- Highlighters (for each student)

- Examples of good and bad swing set designs and locations (digital or paper copies)

- Three images of swing sets (for each team) or preselected links to display to the class

Note: Optional links are located in the Internet Resources section; example swing set images are attached at the end of the lesson.

Additional Materials for Forces Push Back

- 8–10 sets of sports equipment (e.g., jump rope, Frisbee, hacky sack, Wiffle bat and ball, beanbag toss, football, soccer ball, volleyball, tennis racket and ball, basketball)

- 8–10 clipboards

- 8–10 pencils

- 1 timer (for teacher)

- Safety glasses or goggles

Additional Materials for ELA Connection

- The book *Wednesday, A Walk in the Park* by Phylliss DelGreco, Jaclyn Roth, and Kathryn Silverio (JumPsKip Productions LLC, 2011); or you may provide an alternative story about walking in the park

- Materials to decorate the cover page of the STEM Research Notebook (e.g., markers, colored pencils, scissors, glue stick, stickers, paper scraps)

- Optional literature connection for introducing Newton's first law of motion: the book *Newton and Me* by Lynne Mayer (Sylvan Dell Publishing, 2010)

SAFETY NOTES

1. All laboratory occupants must wear safety glasses or goggles during all phases of this inquiry activity.

2. Use caution when working out in the field as there can be several trip/fall or slip/fall hazards (e.g., sports equipment, uneven ground, holes) that can cause physical injuries.

3. Make sure there are no fragile materials in the area where activities are taking place.

4. Have an appropriate level of adult supervision to ensure safe behaviors during activities.

5. Use caution when working with sharps (e.g., scissors, sticks) to avoid cutting or puncturing skin or eyes.

6. Make sure all materials are put away after completing the activity.

7. Wash hands with soap and water after completing this activity.

CONTENT STANDARDS AND KEY VOCABULARY

Table 4.1 lists the content standards from the *Next Generation Science Standards* (*NGSS*), *Common Core State Standards* (*CCSS*), and the Framework for 21st Century Learning that this lesson addresses, and Table 4.2 (p. 51) presents the key vocabulary. Vocabulary terms are provided for both teacher and student use. Teachers may choose to introduce some or all of the terms to students.

Table 4.1. Content Standards Addressed in STEM Road Map Module Lesson 1

NEXT GENERATION SCIENCE STANDARDS

PERFORMANCE EXPECTATIONS
- 3-PS2-1. Plan and conduct an investigation to provide evidence of the effects of balanced and unbalanced forces on the motion of an object.
- 3-PS2-2. Make observations and/or measurements of an object's motion to provide evidence that a pattern can be used to predict future motion.

SCIENCE AND ENGINEERING PRACTICES

Planning and Carrying Out Investigations
- Plan and conduct an investigation collaboratively to produce data to serve as the basis for evidence, using fair tests in which variables are controlled and the number of trials considered.
- Make observations and/or measurements to produce data to serve as the basis for evidence for an explanation of a phenomenon or test a design solution.

Analyzing and Interpreting Data
- Represent data in tables and/or various graphical displays (bar graphs, pictographs and/or pie charts) to reveal patterns that indicate relationships.
- Analyze and interpret data to make sense of phenomena, using logical reasoning, mathematics, and/or computation.
- Compare and contrast data collected by different groups in order to discuss similarities and differences in their findings.
- Analyze data to refine a problem statement or the design of a proposed object, tool, or process.

Continued

Table 4.1. (*continued*)

Using Mathematics and Computational Thinking

- Decide if qualitative or quantitative data are best to determine whether a proposed object or tool meets criteria for success.

- Organize simple data sets to reveal patterns that suggest relationships.

- Describe, measure, estimate, and/or graph quantities (e.g., area, volume, weight, time) to address scientific and engineering questions and problems.

Obtaining, Evaluating, and Communicating Information

- Read and comprehend grade-appropriate complex texts and/or other reliable media to summarize and obtain scientific and technical ideas and describe how they are supported by evidence.

- Compare and/or combine across complex texts and/or other reliable media to support the engagement in other scientific and/or engineering practices.

- Combine information in written text with that contained in corresponding tables, diagrams, and/or charts to support the engagement in other scientific and/or engineering practices.

- Obtain and combine information from books and/or other reliable media to explain phenomena or solutions to a design problem.

- Communicate scientific and/or technical information orally and/or in written formats, including various forms of media as well as tables, diagrams, and charts.

DISCIPLINARY CORE IDEAS

PS2.A: Forces and Motion

- Each force acts on one particular object and has both strength and a direction. An object at rest typically has multiple forces acting on it, but they add to give zero net force on the object. Forces that do not sum to zero can cause changes in the object's speed or direction of motion.

- The patterns of an object's motion in various situations can be observed and measured; when that past motion exhibits a regular pattern, future motion can be predicted from it.

PS2.B: Types of Interactions

- Objects in contact exert forces on each other.

CROSSCUTTING CONCEPTS

Cause and Effect

- Cause and effect relationships are routinely identified, tested, and used to explain change.

Patterns

- Patterns of change can be used to make predictions.

- Patterns can be used as evidence to support an explanation.

Continued

Table 4.1. (*continued*)

Structure and Function
- Different materials have different substructures, which can sometimes be observed.
- Substructures have shapes and parts that serve functions.

Influence of Science, Engineering, and Technology on Society and the Natural World
- People's needs and wants change over time, as do their demands for new and improved technologies.
- Engineers improve existing technologies or develop new ones to increase their benefits, to decrease known risks, and to meet societal demands.

COMMON CORE STATE STANDARDS FOR MATHEMATICS

MATHEMATICAL PRACTICES
- MP1. Make sense of problems and persevere in solving them.
- MP2. Reason abstractly and quantitatively.
- MP4. Model with mathematics.
- MP5. Use appropriate tools strategically.
- MP6. Attend to precision.
- MP7. Look for and make use of structure.

MATHEMATICAL CONTENT
- 3.OA.B.5. Apply properties of operations as strategies to multiply and divide.
- 3.OA.D.9. Identify arithmetic problems (including patterns in the addition table or multiplication table), and explain them using properties of operations.

COMMON CORE STATE STANDARDS FOR ENGLISH LANGUAGE ARTS

READING STANDARDS
- RI.3.5. Use text features and search tools (e.g., key words, sidebars, hyperlinks) to locate information relevant to a given topic efficiently.
- RI.3.7. Use information gained from illustrations (e.g., maps, photographs) and the words in a text to demonstrate understanding of the text (e.g., where, when, why, and how key events occur).
- RI.3.10. By the end of the year, read and comprehend informational texts, including history/ social studies, science, and technical texts, at the high end of the grades 2–3 text complexity band independently and proficiently.

Continued

Table 4.1. (*continued*)

WRITING STANDARDS

- W.3.1. Write opinion pieces on topics or texts, supporting a point of view with reasons.

- W.3.1.D. Provide a concluding statement or section.

- W.3.3. Write narratives to develop real or imagined experiences or events using effective technique, descriptive details, and clear event sequences.

SPEAKING AND LISTENING STANDARDS

- SL.3.4. Report on a topic or text, tell a story, or recount an experience with appropriate facts and relevant, descriptive details, speaking clearly at an understandable pace.

- SL.3.6. Speak in complete sentences when appropriate to task and situation in order to provide requested detail or clarification.

FRAMEWORK FOR 21ST CENTURY LEARNING

- Interdisciplinary Themes: Health and Safety; Environmental Literacy; Science; Mathematics

- Learning and Innovation Skills: Creativity and Innovation; Critical Thinking and Problem Solving; Communication and Collaboration

- Information, Media, and Technology Skills: Information Literacy; Media Literacy

- Life and Career Skills: Flexibility and Adaptability; Initiative and Self-Direction; Social and Cross-Cultural Skills; Productivity and Accountability; Leadership and Responsibility

Table 4.2. Key Vocabulary for Lesson 1

Key Vocabulary	Definition
accelerate	to increase the rate of speed during a given amount of time
actuary	a person who looks at patterns in numbers and data and uses them to make predictions
air resistance	the force that works against a moving object in the air; also called *drag*
balanced force	two forces that are equal in strength and are being pushed or pulled in opposite directions
brainstorm	a group discussion to create ideas and solve problems
direction	the path a moving object follows
feedback	giving input to improve a system or process
force	an action that causes an object to be in motion or change in motion (a push or a pull)
friction	a force that causes moving things to slow down when one object rubs against another object

Continued

Table 4.2. (*continued*)

Key Vocabulary	Definition
geographer	a person who collects data about Earth's landforms and environment to make decisions about wildlife, weather, and climate
gravity	the force that pulls objects toward the center of the Earth
inertia	the property of an object that keeps it resting when at rest or moving when in motion in the same straight line unless acted on by an outside force
Isaac Newton	an English mathematician and physicist who is remembered for his law of gravitation and his three laws of motion
materials	the matter from which something is made
mechanical	work created by machines
mechanical engineer	a person who researches, designs, makes, and tests all kinds of mechanical devices such as machines, engines, and tools
model	a representation of an object that includes the objects' major features; usually a smaller scale of the original
motion	the act of changing place or position
Newton's laws of motion	the three physical laws that describe the forces that act on an object and how they affect its motion
playground	an outdoor area provided for children to play
prototype	a working model used to refine a final design; includes working parts and is often full size
pull	a force that moves something toward a person or object
push	a force that moves something away from a person or object
resistance	something that slows or stops movement or keeps it from happening
speed	how fast or slow an object is moving
swing set	a structure for children to play on that often has one or more swings and a slide
system	a group of parts working together to perform a function
three-dimensional	something that has or seems to have length, width, and height such as a math cube (also known as 3-D)

Continued

Table 4.2. (*continued*)

Key Vocabulary	Definition
two-dimensional	something that has or seems to have length and width, but no height, such as a piece of paper (also known as 2-D)
unbalanced force	two forces that are pushing or pulling in opposite directions and one force is stronger than the other
velocity	how fast an object moves in a given direction
working definition	an explanation of a term or phrase that is formulated for a specific purpose or project

TEACHER BACKGROUND INFORMATION
Engineering and Engineering Careers

In this module, students will act as mechanical engineers who use science and mathematics to design, build, and maintain equipment. Many students have experienced creating their own structures using three-dimensional building blocks; consequently, the students will be familiar with using many of the same decision-making processes that engineers use. Mechanical engineers must understand scientific principles (physics), as well as have an understanding of materials (chemistry). This lesson will help students understand that decisions about each component are made after careful deliberation and testing.

In addition to mechanical engineers, many other kinds of engineers design and build different things:

- Aerospace engineers
- Biomedical engineers
- Chemical engineers
- Civil engineers
- Computer engineers
- Electrical engineers
- Environmental engineers
- Forensic engineers
- Genetic engineers
- Military engineers
- Nuclear engineers

- Reverse engineers

- Software engineers

- Structural engineers

For an overview of the various types of engineering careers, visit any of the following websites:

- *www.aboriginalaccess.ca/adults/types-of-engineering*

- *www.sciencekids.co.nz/sciencefacts/engineering/typesofengineeringjobs.html*

- *www.egfi-k12.org*

- *www.engineergirl.org/9311/What-They-Do*

The "Engineering for Kids" video from 2025Labs also offers an excellent overview on what an engineer is and does: *www.youtube.com/watch?v=UiaXl0giP78.*

Forces and Newton's First Law of Motion

The concept of forces will be introduced as pushes or pulls in this lesson. Students should understand that a force (a push or pull) can cause an object to start moving, to stop moving, or to change the direction of its motion. You may wish to use simple demonstrations, such as rolling a ball and stopping it and rolling a ball and diverting its path by pushing it in a new direction, to reinforce student understanding of forces. The focus of the discussion of forces in this module is on cause and effect, and students should understand that all motion is caused by the application of forces. The forces that cause motion are known as unbalanced forces. An unbalanced force is when two or more forces act on an object unequally (i.e., they are not balanced). An object's motion will not change unless an unbalanced force acts on it. You may wish to demonstrate this to students by pushing on a wall with your hands and asking students if the wall moved, explaining that this is an example of balanced forces with two forces or pushes (your hands pushing against the wall and the wall pushing back against your hands) balancing each other so that there is no motion. In contrast, if you push a chair with your hands it will move because the forces are unbalanced. Students should understand that unbalanced pushes and pulls cause motion.

This lesson also introduces Newton's first law of motion, also known as the law of inertia, which states that an object in motion will stay in motion with the same speed and moving in the same direction unless acted on by an outside (unbalanced) force, and an object at rest will stay at rest unless acted on by an outside (unbalanced) force. This lesson will also provide opportunities for students to explore the motion of objects and the forces that cause changes in motion, including gravity and friction.

There are many everyday examples of Newton's first law of motion. For example, to move a box across the floor you must push or pull it. Using students' experiences riding in cars can provide useful examples. For example, you often use covered cups for beverages while riding in a car because liquid in an uncovered cup tends to spill because of inertia. The water or other liquid in the cup will tend to stay at rest when the driver starts driving forward. When the car accelerates, the water tends to remain at rest, the car moves out from under it, and the water sloshes toward the person holding it. On the other hand, when the driver brakes a moving car, the water continues moving forward with the same speed and in the same direction as the car was moving, and will slosh forward in the cup when the car stops.

For additional information about force and motion generally and in the context of sports activities, see the following resources:

- *http://studyjams.scholastic.com/studyjams/jams/science/index.htm?topic_id=fnm*

- *www.scientificamerican.com/article/bring-science-home-frisbee-aerodynamics*

- *www.real-world-physics-problems.com/physics-of-sports.html*

Playgrounds and Parks

Students will view images of several playgrounds and swing sets during this lesson as they evaluate factors that make playgrounds fun. Images are provided at the end of this lesson; you may choose to provide additional images that reflect the types of playground equipment that may be found in your setting (e.g., urban, rural). Different settings provide different challenges for the placement and type of playground equipment. You may wish to point out features of these settings to students by discussing, for example, how swing sets in rural areas may differ from those in urban settings (e.g., space limitations influence the types of equipment available).

Students will also consider playground safety throughout the module and will compile a class list of safety considerations. If your school or public playground has a set of rules for safe use of playground equipment, you may wish to have this on hand to facilitate the discussion. You may also wish to compile a list of safety features present in your school's playground (e.g., types of mulch that cushion falls, coated metal surfaces to prevent burns from heated metal).

Students will explore local, state, and national parks throughout the module. The following websites may be helpful to review for information about public parks. You may wish to direct students to these resources as they conduct research for the Park Exhibition Showcase activity (see handout on p. 87):

- *www.nrpa.org/publications-research*

- *www.stateparks.org/find-a-park*

- *www.tpl.org/center-city-park-excellence*
- *www.rff.org/files/sharepoint/WorkImages/Download/RFF-BCK-ORRG_Local%20Parks.pdf*
- *www.everykidinapark.gov*
- *www.discovertheforest.org/?m=1#map*

Civic Responsibility

Citizenship means being a member of and supporting one's community. In the United States, citizens have been guaranteed a number of freedoms under the Bill of Rights. With these freedoms, however, comes the responsibility to be informed, uphold the law, be tolerant of others, and participate in local projects dedicated to the common good.

In her blog, 20 Ideas for Teaching Citizenship to Children (*www.kellybear.com/TeacherArticles/TeacherTip27.html*), Leah Davies gives practical pointers for adding citizenship content to curriculum seamlessly. Having open discussions, inviting guest speakers, writing stories and poems, reading, and debating are a few of the recommended tactics.

COMMON MISCONCEPTIONS

Students will have various types of prior knowledge about the concepts introduced in this lesson. Table 4.3 outlines some common misconceptions students may have concerning these concepts. Because of the breadth of students' experiences, it is not possible to anticipate every misconception that students may bring as they approach this lesson. Incorrect or inaccurate prior understanding of concepts can influence student learning in the future, however, so it is important to be alert to misconceptions such as those presented in the table.

Table 4.3. Common Misconceptions About the Concepts in Lesson 1

Topic	Student Misconception	Explanation
Forces	Sustaining motion requires a continued force.	Newton's first law of motion declares that a force is NOT needed to keep an object in motion. For example, consider a pencil rolling on a desk. An object comes to a state of rest because of the presence of an outside force such as friction or air resistance acting on the object.
	Everything that moves will eventually come to a stop. Rest is the "natural" state of all objects.	Newton's first law of motion also debunks this myth. Everything at rest will stay at rest, and everything in motion will keep moving until an external force acts upon the objects.

PREPARATION FOR LESSON 1

Review the Teacher Background Information provided (p. 53), assemble the materials for the lesson, make copies of the student handouts, and preview the videos recommended in the Learning Components section that follows. Have your students set up their STEM Research Notebooks (see pp. 25–26 for discussion and student instruction handout).

Hang chart paper and have markers on hand for creating a KWL chart. Make copies of three swing set images for each team to evaluate or use preset saved images from websites (possible images are attached at the end of this lesson on p. 76 and suggested websites are provided in the Internet Resources section, p. 74). Set up the sport stations for science (with one sport activity at each station), and preview the links and videos in the lesson plan. Generate a list of your local parks for teams to select one from for research and exhibition.

An optional literature connection for introducing Newton's first law of motion is *Newton and Me* by Lynne Mayer. You may wish to have this book on hand and others that show people or objects being moved by a variety of forces as well as other nonfiction books about force (see Table 4.4, p. 58, for examples). You may also wish to have books available that feature parks and playgrounds in various settings.

Table 4.4. Nonfiction Books About Force

Book	Author	Publisher, Year
And Everyone Shouted, "Pull!" A First Look at Forces and Motion	Claire Llewellyn	Picture Window Books, 2004
Forces and Motion: From High-Speed Jets to Wind-Up Toys	Tom DeRosa and Carolyn Reeves	Master Books, 2009
Forces Make Things Move	Kimberly Brubaker Bradley	HarperCollins, 2005
Give It a Push! Give It a Pull! A Look at Forces	Jennifer Boothroyd	Lerner Publishing Group, 2010
Gravity	Jason Chin	Roaring Brook Press, 2014
Gravity Is a Mystery	Franklyn M. Branley	HarperCollins, 2007
I Fall Down	Vicki Cobb	HarperCollins, 2004
The Magic School Bus Plays Ball: A Book About Forces	Joanna Cole	Scholastic, 1998
Why Do Moving Objects Slow Down? A Look at Friction	Jennifer Boothroyd	Lerner Publishing Group, 2010

Students will be working in groups, called design teams, throughout the module. Be prepared to divide students into groups of about four students each before passing out the swing set images during the introduction to the lesson.

LEARNING COMPONENTS
Introductory Activity/Engagement

Connection to the Challenge: Begin the lesson by introducing the module challenge: Student design teams will be challenged to survey existing playground equipment, compare and contrast different designs in light of safety concerns connected to the playground equipment and its environment, and then sketch and build a small-scale model of their proposed design using geometric shapes and measurements. Details of the swing set and evidence of its improvement on existing swing sets will be synthesized in a blog post.

Students will create swing set designs that

- include a swing, a slide, and some sort of connecting partition;

- take advantage of features that maximize lessons about force and motion; and

- use fun features outlined in discussions while maintaining high standards of safety.

Introduce the term *model*. Have students share their ideas of what a model is, leading students to a definition of a model as a representation of something that is usually smaller than the actual object. Hold a class discussion about why models might be useful (e.g., allows us to see the working parts of something that is too large to work with easily; allows us to experiment with features of an object without making changes to the object itself). Next, ask students if they have heard of the term *prototype*. Tell students that engineers often use a type of model that contains working parts to refine a final design, and it is often full size; this is called a prototype. Ask students to share their ideas about whether their models will be prototypes (e.g., the models will contain some functioning parts, such as the swing; however, the models will be smaller than an actual swing set and will not be constructed of the materials that would be used as an actual swing set—so they are not true prototypes according to this definition).

Begin each day of this lesson by directing students' attention to the driving question for the module: How can I use what I know about force and motion to create a plan and build a model of a swing set that is both fun and safe? Remind students that their challenge is to plan and build a model for swing set components that meets criteria for fun and safety. Review the driving questions for Lesson 1 (see below) and briefly discuss how the students' learning in the previous days' lesson(s) contributed to their ability to create their plan and build their prototype.

Driving Questions for Lesson 1: What forces interact with your body as you play on the playground? How do these forces affect your experience of fun (the "fun factor") when you play on the swing set?

Ask students to share their experiences on playground or park swing sets. Possible discussion questions include the following:

- Do you play on the swing set on the school playground or neighborhood park?

- What playground equipment do you enjoy most?

- Fill in the blank: To be fun, a swing set must have _____.

- What makes the different parts of a swing set move?

- Why do you slide down the slide on a swing set and don't slide up?

- Why does your swing stop at the top and then seem to drop and swing back?

- What is that pull you feel when you are climbing up a rock wall or a ladder?

Science Class: Introduce the STEM Research Notebooks to students using the STEM Research Notebook guidelines found on page 26. Review the STEM Research Notebook Rubric (p. 88) with students and have them tape or glue a copy to the inside covers of their notebooks.

Ask students to create a KWL chart in their STEM Research Notebooks for swing sets. Next, ask them to respond to these questions in their charts:

- What do you know about swing sets?

- Is there anything that you want to know or are curious about? About the forces that affect motion on the swing set? About the design of a swing set? Are you curious about the shapes used for parts of the swing set or why ropes or chains are used and not a pole for the swing? Is it possible to go up and over the bar of a swing set or will gravity and other forces prevent this from occurring? What other things might you be curious about?

- What have you learned so far? (Students should begin generating the final column, but remind them that this is a work in action. Provide students with an opportunity each day throughout the module to make modifications to this chart.)

Show several images of swing sets with some that are poorly placed or have a poor design included. Students can hold up signs they create to indicate designs that are a win or a fail. The following websites contain images of multiple swing sets:

- *www.swingsets.com/swing-sets*

- *www.overstock.com/Sports-Toys/Swing-Sets/19670/subcat.html*

- *www.slideinnovations.com*

- *www.homedepot.com/b/Playsets-Recreation-Parks-Playsets-Playhouses-Playsets-Swing-Sets/N-5yc1vZc5p2*

- *www.backyardadventures.com*

Group students into design teams of four students each. Ask design teams to share their concerns about these playground swing sets and to answer the following questions:

- Is this swing set fun? What qualities make this swing set fun or not fun?

- Is this swing set safe? What conditions cause this swing set to be unsafe?

Have students go back to add to the KWL chart if these images have sparked new ideas about what they know about forces.

Mathematics Connection: Not applicable.

ELA Connection: Introduce the book *Wednesday, A Walk in the Park* by Phyllis DelGreco, Jaclyn Roth, and Kathryn Silverio, with the following questions:

- What comes to your mind when you hear the words, "Let's go to the park?"

- What can you do at the park that you cannot do at home in your house?

- If you were a grown-up taking a child to the park, what would you want them to do there?

Read the book *Wednesday, A Walk in the Park.* This is a story about a child named Jessie spending time in the park with her grandpa as he imparts his wisdom. Possible reading response questions include the following:

- What season of the year did Jessie and Grandpa visit the park? How do you know?

- What wisdom did Grandpa impart when he saw the children arguing over the candy?

- What did he have to share when the child fell down?

- Why do you think the author wrote about these incidents?

- What elements in this book helped you recognize that this is a fiction book?

- How could this story be changed to make it a nonfiction book?

Social Studies Connection: Introduce the concept of safety. Ask students to work as a class to brainstorm ideas of things they encounter in their everyday lives that are designed to keep them safe (e.g., bike helmets, seat belts, car air bags). Ask students to share their ideas about features of playgrounds that are designed to help them play safely (e.g., age recommendations for swing sets, materials that won't get hot in the sun, mulch and other ground materials that will cushion falls). Create a class list of students' ideas. Ask students if they have seen a list of rules at a playground they have visited. Have students share some ideas about what rules for behavior might help ensure safety on the playground, particularly with regard to swings and slides.

Activity/Exploration

Science Class: Introduce the Fun Factor Survey (p. 77) by asking students to share their ideas about how they can predict whether playground equipment will be fun (e.g., what features do they look for?). Introduce the idea of measurement using the following discussion prompts and questions:

- Scientists often measure things to learn how objects and energy interact. Some things are easy to measure, such as the size of a book, the length of your pencil, or

the mass of a baseball. Can you think of things that are hard to measure? (Guide the responses toward intangible things such as love, fun, or happiness.)

- Pretend our class has been chosen by a game company to rate some new games their designers are working on. The company is supplying the class with early versions of the games. The designers want to build their games so that kids like you want to play them. What kind of information can we give the designers that will be helpful? How will we get that information?

- How might this relate to our Swing Set Makeover Design Challenge? What information will you need to consider as you design and provide feedback for the swing sets?

Pass out the Fun Factor Survey. The survey will help students identify characteristics of fun that can apply to swing set designs. Allow each student to assess his or her own perception of "fun" by filling out the survey to establish a baseline measurement. Students should record their personal preferences for fun in the first chart by numbering the squares 1 to 5. Survey components include the following:

- Speed: Does the student like to go fast (not at all or moving as fast as a cheetah)?

- Flight: Does the student want to fly through the sky? (A penguin cannot fly, a baby bird hops when it tries—but an eagle soars!)

- Height: Does the student like heights (being up in the clouds) or prefer to stay close the ground?

- Spin: Some students do not like spinning; others love to spin around and around.

- Fun Spaces: These are the spaces the student enjoys most—wide open spaces, spaces with puzzles or gadgets, or interconnected spaces forming a maze.

Now that each student has identified her or his own personal standards for fun, tell the students they will work in their design teams to begin evaluating swing set designs and imagining their own designs for swing sets that will be fun and safe. Students will remain in these design teams for the rest of the module. Encourage team members to give their team a name. Next, students should share their survey results with their team members. Ask students to answer the following questions about their team members' individual rubrics: Are all the items numbered the same on each survey? How are they similar? How are they different?

Explain to students that they will create a common rubric based on the team members' survey results. Let teams brainstorm to devise a strategy for selecting the order of the items on the survey.

Present swing set #1 either from among the images provided at the end of this lesson (p. 76) or one that was selected during the lesson preparation, and project it on a Smart Board. Walk students through the process of evaluating it using their teams' fun factor rubrics. Allow time to add up the score and propose a suggestion for improving the swing set design. Teams will evaluate swing sets #2 and #3 in the same manner. Be sure to provide students with a variety of swing set designs to assess to allow students to ensure that their rubrics provide the evidence they need to determine the swing set's fun factor.

Take the class on a "field trip" to the school playground. Teams will complete their rubrics to evaluate the school playground swing set(s). They will also make suggestions for improvement to the school playground based on their fun factor scores.

Introduce the force of gravity by drawing a picture of the path a basketball takes when you shoot a basket on the playground basketball court on Earth. Next, tell students that you will pretend there is a basketball court on the International Space Station (a low gravity environment). Draw the path of the basketball in this low gravity situation. Ask students to predict which basketball will travel farther and explain their predictions. (*Note:* The example of the space station is used rather than the moon or another planet to illustrate the influence of gravity; the moon and other planets also exert gravitational forces on objects, so the ball in these environments would be under the influence of gravity and therefore the path of a thrown ball on the moon or another planet will be similar with the path on Earth, but with a steeper or less steep arc.)

Ask students to share their ideas on what forces are acting on the ball on Earth versus on the space station. Introduce the concept of gravity.

Next, ask students to consider what force keeps a basketball player from sliding across the floor without stopping (e.g., compare the basketball player's movement to an ice skater's). Introduce the concept of friction. Ask students to share their ideas about how friction aids a basketball player on the court.

Next, place a soccer ball on the floor and stare at it. Announce to students that you are playing soccer. Ask students whether they agree or disagree with your statement. Ask students to share their ideas about what else you need to do to play soccer. Students should associate playing soccer with motion. Introduce Newton's first law of motion (see Teacher Background Information, p. 54), guiding students to understand that motion is the result of forces that push and pull. Explain to students that the forces of gravity and friction also play a role in the player's and the ball's motion in soccer. Demonstrate the following concepts to the class:

- *Inertia.* Balls in motion tend to keep moving in a straight path unless acted on by an outside force. (Push the ball across the room and ask students for their ideas about why it eventually stops; introduce the idea that if there were no forces acting on the ball that it would keep rolling indefinitely.)

- *Gravity.* This is the downward pull that keeps things on the ground or causes objects to fall to the ground. (Toss a ball and ask students for their ideas about why it falls back toward the ground.)

- *Friction.* This is the rubbing force that causes objects to slow down when in motion. (Ask students how the motion of the ball might be different if it was rolled on a smooth sheet of ice; introduce the idea that surfaces rubbing against one another create friction.)

Show students a video about forces and gravity such as "Gravity, Force, and Work" found at *www.youtube.com/watch?v=LEs9J2IQIZY*. After watching the video, have students work in teams of three or four to develop definitions for the terms *motion, force,* and *gravity,* including examples. Lead a class discussion in which teams share their definitions and support their definitions with examples.

STEM Research Notebook Prompt

Write definitions of the words force, gravity, *and* friction *in your STEM Research Notebook. Next, draw a picture of you playing your favorite sport. Label the forces in your picture.*

Forces Push Back

Students will engage in the Forces Push Back activity using what they have learned about inertia, forces, and motion to identify the push and pull forces in the playground and sports stations you have set up around the room or playground.

1. Before class, set up 8 to 10 stations (one for each team of three or four students) that represent a different sporting event or playground activity. These can be set up in the gym, on the playground, or in a large classroom.

2. Have each team collect the materials necessary for the activity.

3. Pass out the Forces Push Back handout (p. 85) and review the following procedures with the class to ensure students know what to do:

   - Write the name of the sport in the table.

   - Use the equipment to play the sport for three minutes.

   - While playing, pay attention to the forces used to play.

   - Circle on the handout if a push or pull force was being used, or both, and explain your reasoning. For example, a baseball is a push because your hand and the bat push on the ball to make it move.

- Complete the sentence threads to provide examples of inertia and motion.

- Circle the outside force(s) you identified (gravity or friction).

- Go to the next station. (It is helpful if the stations are numbered to prevent confusion.) *Note:* There should be an agreed signal that will indicate that it is time to move on to the next station.

4. Send the students to the stations and use the timer to allow three minutes for exploring. Then use a signal (ring a bell or blow a whistle if you are outside) to get students to stop playing and begin writing about their experiences. Allow one minute for writing.

5. Signal to indicate that students move to the next station and begin again until everyone has been to every station.

6. When each group has been to every station, clean up and have students return to their seats.

7. Chart class findings on the board and discuss the results. Questions may include the following:

- What observations did you make about the role of inertia in the sports you played?

- What observations did you make about the role of gravity and friction in the sports you played?

- How can you apply this information to playing on a swing set?

- How will what you know about inertia, gravity, and friction affect your design decisions? Write a two-paragraph summary in your STEM Research Notebook. Describe your thoughts and "aha" moments during the investigations in paragraph one. Explain your conclusions and use evidence to support them in paragraph two.

8. Allow 15 minutes for students write their reflections in their STEM Research Notebook.

Bar Graph

Mathematics Connection: Graph the Fun Factor Survey scores on a bar graph. A handout is provided for students who need help setting it up (Fun Factor Bar Graph Template, p. 84). Each student should do the following:

- Add all the scores for each swing set to get a total fun factor score.

- Create a bar graph with fun factor values (5–25) on one axis and the labels for each set of data on the other.

Hold a class discussion asking for student ideas about how mathematics can be used to help make decisions. Guide students to analyze the results of their data to determine which swing set is the most fun according to their data.

STEM Research Notebook Prompt

Have students write a paragraph summary (3–5 sentences) about their Fun Factor Surveys. You may wish to give students the following outline to use:

- Sentence 1: What did I do?

- Sentence 2: Why did I do this?

- Sentence 3: What did I find out?

- Sentence 4: How can I apply what I learned?

Many careers involve mathematics, including game developers, animators, robot engineers, and doctors and nurses. Career connections you may want to introduce that relate to the mathematics content of this lesson include the following:

- Statisticians collect data and study and analyze the data. They can use these data to help companies develop good business strategies.

- Geographers collect data about the Earth's landforms and environment and use the data to make decisions about the effect the environment has on wildlife, weather, and climate patterns.

- Actuaries look at patterns in numerical data and apply theories to make predictions.

- Stockbrokers use mathematics to guide people who want to invest their money.

ELA Connection: Students may design a cover page for their STEM Research Notebooks using the title "Swing Set Makeover." Allow students to use art materials, add pictures they have printed, word art, or hand-drawn images. The cover should show what part of this module excites the student most. Students should add the title to the table of contents on the first page of their STEM Research Notebooks and continue numbering the next 10 pages.

Note: You may want to use this opportunity to launch a discussion about tables of contents, emphasizing that these are typically found in nonfiction texts.

Park Exhibition Showcase

Social Studies Connection: Introduce the Park Exhibition Showcase activity by holding a class discussion about parks, asking questions such as the following:

- Why do cities provide park spaces in the community? (Possible answers include that it is healthy, to provide a place for the community to relax, or to provide a space for water.)

- Who is responsible for making park spaces in a community? (Make sure students understand that city and other local governments provide local parks.)

- What are some of the needs of our community park spaces?

- Who takes care of parks and what kind of work do they need to do to care for the parks? (Typically, a parks board cares for public parks; students should understand that taking care of all of the parks in a city is a big job. Students may be able to relate to the enormity of the task by considering the amount of work it takes to care for their own yards.)

Write the names of several local parks in a column on the board.

- Take a poll to see how many students have visited each of the parks listed on the board.

- Record the totals beside each park name.

- Ask students to create bar graphs in their STEM Research Notebook to show how many classmates have been to each of the parks listed on the board.

- Ask, "Are all of the parks the same or are some parks better than others?"

- Ask students to share their ideas about why they think some parks are better than others.

Some cities have exhibitions for their parks to showcase special attractions and to pique the public's interest in particular parks. Student teams will research parks and participate in a Park Exhibition Showcase:

- Student teams will each choose a park that they want to showcase.

- Teams will each prepare an exhibit to put on display in the school library to share the wonder of the city parks with their friends and peers.

Pass out and read through the Park Exhibition Showcase handout (p. 87). A student from each team should be prepared to visit the team's park to take pictures and make observations about park swing sets and play areas. You may also wish to take this

opportunity to invite your city park commissioner or a local park staff person to come to your class to talk about their job and how citizens contribute to and help care for city parks.

Ask students to evaluate public involvement in helping with the upkeep of the park they are studying:

- Based on evidence from the internet and personal observations, does the community help take care of this park or contribute to the park in any way (e.g., donate park benches, trees, help care for the gardens)?

- Is it a citizen's responsibility to help care for local parks and other public properties? (Have students debate this question.)

Explanation

Science Class: Continue the discussion about forces you began in the Activity/Exploration section and relate what students learned about forces to swing sets. Hold a class discussion about swing sets and forces, asking questions such as the following:

- What kind of actions do you do on a swing set? (Possible answers include pumping feet, climbing ladder, and sliding.)

- Where are forces applied on a swing set? (Students should conclude that forces cause motion—going up, going down, spinning, bumping—all result from a force being applied.)

Explain to students that swing sets allow the rider to experience height and speed due to the forces that push and pull on their bodies. Swing sets allow riders to test their strength and endurance as they work with and against gravitational forces. Explore the idea that a swing set offers opportunities for falling in a controlled manner; gravity is an important consideration when planning swing set structures.

Models. Students should understand that models are small-scale versions of a product that can be used for testing so that improvements can be made. Students will be making annotated sketches throughout the module, which they may use to build a model in Lesson 4.

Careers. The career focus for this module will be on students' roles as mechanical engineers. You may wish to read books about men and women mechanical engineers and weave in online videos showing the fascinating work mechanical engineers do.

Fun Factor Survey. Being self-aware and flexible are key learning goals for the Fun Factor Survey activity. The reason students do a self-assessment first is to help each student become self-aware before being influenced by peers. This self-assessment will also be considered when students build the swing set model later. If a student in his or her self-assessment indicates he or she prefers swing sets with high places and this is not part of

the team's goal, the reason for its omission should be reflected when he or she writes his or her final reflection.

Forces. Force is anything that tends to change the state of rest or motion of an object. Forces cause changes in the speed or direction of the motion of an object; the greater the force placed on an object, the greater the change in motion. The more massive an object is, the less effect a given force will have on its motion. The activities in this lesson are based on a working definition for force. A *working definition* is one that is determined by students. It may or may not be completely accurate; however, it should be used and corrected by the students as they gain more experience with and understanding of the concept. The advantage of a working definition is that it is an indicator of student under-standing that you can use to guide future experiences.

Explain that there are invisible forces including gravity and friction. They are like the wind. We can't see these forces, but we know they are there because we can see their effect. We can't see gravity, but we know it is there because the natural tendency of objects is to fall down. We can't see friction, but we know it is there by watching the way things move. Friction helps things stay in one place.

Tables 4.5 and 4.6 (pp. 70–71) contain information that may be useful for the Forces Push Back activity.

Table 4.5. Playground Sports and Activities That Push or Pull

Playground Sports and Activities That Push	
Baseball/Racquetball	Hand pushes ball, bat/racquet pushes ball
• Inertia causes the moving ball to move in a straight line until a person's hand/the wall or the bat/racquet stops it or air pressure slows it down (friction) and gravity pulls it to the ground.	
Kickball/Soccer	Foot pushes ball
• Inertia causes the moving ball to move in a straight line until a person's foot causes it to change direction (push) or a person's arms or friction (from the ground) and gravity slow it down.	
Football	Hand pushes ball, foot pushes ball
• Inertia causes the moving ball to move in a straight line until another force or forces slows or stops it; this force may be the force exerted by a person's arms when he or she catches the ball, the force of gravity that results in the ball hitting the ground, or frictional forces such as air pressure that slow the ball's motion.	
Basketball	Hand pushes ball
• Inertia causes the moving ball to move in a straight line until gravity pulls it down into the basket or the ball hits something that deflects it somewhere else. • Force is applied to push the ball to the ground and it bounces back up as the result of the elasticity of the basketball material (dribbling). • A stacked balls demonstration can be shown to explain elastic force to the curious students using a demonstration such as that depicted in a Physics Girl video (*http://thekidshouldseethis.com/post/physics-girl-the-stacked-ball-drop-and-supernovas*).	
Frisbee	Hand pushes Frisbee
• Inertia causes the moving Frisbee to move in a straight line until the air slows it down (friction) and lets gravity take over, pulling it to the ground. Alternatively, the Frisbee may be stopped by a hand or other body part. • A Frisbee simultaneously experiences lift (like an airplane) and spin. Curious students may wish to view the video "Why Does a Frisbee Fly?" (*www.youtube.com/watch?v=IJPaVi2Fxhc*) for optional additional information.	
Hacky sack	Foot pushes sack
• The hacky sack is constantly under the influence of gravity pulling it to the ground, but the force exerted by a body part pushing it up can temporarily overcome gravity.	
Playground Sports and Activities That Pull	
Tug of war	Hand pulls on rope
• Two groups are pulling, which is a force in itself, but gravity is helping, too, as people use their body weight to exert more pull; friction comes into play as well since it prevents feet from moving—or slows them down.	
Sliding down a slide	Rider slides down
• Gravity causes a rider to move from the top of the slide to the bottom	

Table 4.6. Playground Sports and Activities That Push and Pull

Jump Rope	Hand pulls on rope to make it swing; feet push up off the ground to jump

- The rope is being pulled as the jumper swings it around and gravity is always exerting a force (as noted when the jumper slows down the swing).
- Friction is involved as the rope moves through the air (you can hear its whistle). Friction also affects the rope as it hits the ground when the jumper's arms are extended downward, thus slowing it down.
- A body that jumps into the air will always come back down because gravity is pulling on it (on Earth).

Climbing (monkey bars/ladders)	Hand pulls on bar; foot pushes on ground or another bar

- The body overcomes gravity as it pulls on the bar and pushes up with the feet. Friction is occurring where the climber's shoes touch the bar; the tread on the shoe prevents the foot from slipping on the bar as the climber pushes off.

Archery	Hand pulls string; string pushes arrow

- Inertia makes the arrow move in a straight line at a constant speed until friction through the air slows it down and gravity begins pulling it down to the ground … or the arrow strikes something and it stops.

Swimming	Hand pulls body; kicking feet push body

- The swimmer is pushing and pulling to create the forward motion and inertia keeps the swimmer moving forward. Even with buoyancy force acting to keep the swimmer afloat, gravity is pulling the body down (some bodies more than others) and the water is creating friction as it rubs against the body.
- The effect of friction on speed is significant enough that swimmers will sometimes shave the hair off their bodies to gain speed and purchase special swimming wear (caps and suits) to reduce friction – there are some garments that cause a swimmer to be disqualified from an event because they offer the swimmer too much advantage.

Hockey	Hand pushes stick; stick pushes puck

- Inertia causes the puck to keep moving forward in a straight line from when it is struck until a stick stops it or pushes it in a different direction. Friction occurs as the puck rubs across the ice, causing it to slow down. Gravity's pull keeps the puck on the ice.

Mathematics Connection: The steps in creating a bar graph are emphasized in the Bar Graph Rubric, but it may be helpful to brainstorm the parts of a bar graph with students before beginning and record them on a large sheet of paper to display throughout the module. The steps that the students should complete include the following:

- Decide on a title for the graph.
- Draw the vertical and horizontal axes.
- Label the horizontal axes (swing set scores).
- Label the vertical axes (number of points).

- Decide on the scale. Explain that students should consider the least and the greatest number shown on the graph. Discuss what range of numbers should be shown on this bar graph (e.g., begin at five, the lowest score a swing set could receive, and count by fives to 25, the highest score a swing set can receive).

- Draw a bar to show the total for each swing set and label the bars with the names of the swing sets.

ELA Connection: As students conduct park research on the internet, go over the rules for internet safety and review strategies for searching for information on the park websites (e.g., use keywords, look at headings on the sidebars, and follow hyperlinks to locate relevant information; see information in Social Studies Connection section below). Explain the importance of using images and their captions to glean valuable context clues as well. Provide students with some basic guidelines on conducting internet searches for information. First, direct them to use a safe and reliable search engine (if your school uses a particular kid-friendly search engine, direct students there), then encourage them to use the following steps:

1. Enter search terms or keywords (this can be one or more) and search for these terms.

2. Skim the list of sites that appears.

3. Preview sites that you think may be helpful by going to the site and quickly reviewing the information that is there.

4. Spend your time on sites that your preview shows may be helpful.

5. Focus on credible sites (government websites, educational websites from known sources) instead of blogs and sources that you don't recognize.

6. Try using different search terms if your first search didn't lead you to the information you want.

Social Studies Connection: The goal of the Park Exhibition Showcase is for students to experience park playgrounds and swing sets in active ways in addition to experiencing them through images on a screen or handouts. This activity helps students experience change over time in a personal way while gathering information to inform them of the need for a new swing set design. This activity also will allow students to see firsthand that there are members of the community who take an active interest in public areas such as parks.

Elaboration/Application of Knowledge

Science Class: Challenge students to look for applications of inertia, gravity, and friction interactions in other activities, such as picking up a book, carrying an item from one point to another, or riding in a car.

Mathematics Connection: After students have created a bar graph with their team's data, they can create a line plot with each group's data to determine a class mean score for each swing set design.

ELA Connection: Create a section in the STEM Research Notebook to write about the books the class reads about force and motion, swing set design, and engineers. Have students write the name of the book and indicate if the book is fiction or nonfiction, then respond to a reading response question. Examples of reading response questions for fiction and nonfiction include the following:

Fiction

- Why do you think that the author thought this was an important story to tell?

- What real-life situations or events are you reminded of by this story? What characters or situations in the book remind you of those?

- What character from the book would you like to have as a real-life friend? Why?

- Choose one character and describe the problem this character faces in the story. How does he or she solve this problem? What challenges does he or she face along the way?

Nonfiction

- Write a summary of what you learned about force and motion from this book.

- What information did you find interesting or surprising in this book?

- How could the author turn this nonfiction book into a fictional story? What story would the book tell?

- What is the author trying to do in this book?

- How do the pictures or graphics in the book help the author communicate information?

Social Studies Connection: Holding a discussion about citizen responsibility in helping care for local parks can provide an opening for considering other ways citizens care for the community and the civic duty of citizens to lend a hand when things can be done (e.g., picking up paper floating by on the street, keeping sidewalks clear of debris, not riding bikes through people's yards).

Evaluation/Assessment

Students may be assessed on the following performance tasks and other measures listed.

Performance Tasks

- Fun Factor Survey

- Forces Push Back handout

- Bar Graph

Other Measures

- Engagement in class activities and discussions

- Involvement in group work and discussions

- STEM Research Notebook entries

INTERNET RESOURCES

Information about various types of engineering careers

- *www.aboriginalaccess.ca/adults/types-of-engineering*

- *www.sciencekids.co.nz/sciencefacts/engineering/typesofengineeringjobs.html*

- *www.egfi-k12.org*

- *www.engineergirl.org/9311/What-They-Do*

- *www.youtube.com/watch?v=UiaXl0giP78*

Information about public parks

- *www.nrpa.org/publications-research*

- *www.stateparks.org/find-a-park*

- *www.tpl.org/center-city-park-excellence*

- *www.rff.org/files/sharepoint/WorkImages/Download/RFF-BCK-ORRG_Local%20Parks.pdf*

- *www.everykidinapark.gov*

- *www.discovertheforest.org/?m=1#map*

Blog on teaching citizenship

- *www.kellybear.com/TeacherArticles/TeacherTip27.html*

Images of swing sets

- *www.swingsets.com/swing-sets*

- *www.overstock.com/Sports-Toys/Swing-Sets/19670/subcat.html*

- *www.slideinnovations.com*

- *www.homedepot.com/b/Playsets-Recreation-Parks-Playsets-Playhouses-Playsets-Swing-Sets/N-5yc1vZc5p2*

- *www.backyardadventures.com*

Information about force and motion

- *http://studyjams.scholastic.com/studyjams/jams/science/index.htm?topic_id=fnm*

- *www.scientificamerican.com/article/bring-science-home-frisbee-aerodynamics*

- *www.real-world-physics-problems.com/physics-of-sports.html*

Videos about force, motion, and gravity

- *www.youtube.com/watch?v=LEs9J2IQIZY*

- *http://thekidshouldseethis.com/post/physics-girl-the-stacked-ball-drop-and-supernovas*

- *www.youtube.com/watch?v=IJPaVi2Fxhc*

PLAYGROUND SWING SET IMAGES

Name: _____   Team Name: _____

FUN FACTOR SURVEY

Rate how you feel about each standard using the numbers 1 (not fun) to 5 (most fun). This will help you understand what aspects of playground equipment are most and least important to you. After you have rated each standard, circle the most fun aspects of the playground (those that you rated a 5) and draw an X through the least fun aspects of the playground (those that you rated a 1). For example, if you like to move quickly but not extremely fast, your speed standard may look like this:

SPEED	Nope, not me **(sloth)**	Move at a walk, slow and easy **(panda)**	Move at a quicker pace **(giraffe)**	Move fast, but not scary fast **(gazelle)**	Move so fast my heart races **(cheetah)**
Rate 1–5	3	4	5	2	1

When you are done, you will compare your survey with those of others on your team.

MY PERSONAL METRIC FOR FUN

SPEED	Nope, not me **(sloth)**	Move at a walk, slow and easy **(panda)**	Move at a quicker pace **(giraffe)**	Move fast, but not scary fast **(gazelle)**	Move so fast my heart races **(cheetah)**
Rate 1–5					
FLIGHT	Stay grounded like a **penguin**	Hop like a **grasshopper**	Fly like a **chicken**	Fly like a **bumblebee**	Soar like an **eagle**
Rate 1–5					
HEIGHT	Best on the **ground**	As high as a **step stool**	As high as a **bunk bed**	As high as a **telephone pole**	High up in the **clouds**
Rate 1–5					
SPIN	**Nope,** not me	Turn from **front to back** (1/2 turn)	Turn around in a **full circle** (1 turn)	Spin around **in a circle a few times**	Spin around and **around many times**
Rate 1–5					
FUN SPACES	**Wide-open** spaces	Spaces with **windows and doors**	**Tunnels**	Spaces with **gadgets and puzzles**	**Mazes**
Rate 1–5					

Name: _____   Team Name: _____

STUDENT HANDOUT, PAGE 2

FUN FACTOR SURVEY

Did everyone on your team rate each standard the same way?

Why are your ratings so similar or so different?

Your team will create a rubric that will be used to measure the "fun factor" of playgrounds. For example, you may decide that being able to spin around and around is very important to your team, so you would assign "around many times" a high score in your rubric, but turning front to back is not as important, so you would assign "front to back" a lower score on your rubric.

Your team will need to agree on a method for deciding what features you most like and least like as a team. How will you do this?

Name: _____   Team Name: _____

FUN FACTOR SURVEY

Your team must agree on what the most fun and least fun factors are for a playground and create a new rubric that your team will use to score pictures of playgrounds your teacher provides. Once your team has agreed on the most fun and least fun activities for each standard, use the **bold** key words from the My Personal Metric for Fun table and put them in the matching column and assign the other key words a rating to create your own rubric for fun factors. For example, if your team decided that the most fun playgrounds give you opportunities to go fast, but not too fast, your rubric might look like this:

Standard	1	2	3	4	5
Speed	**sloth**	**cheetah**	**panda**	**giraffe**	**gazelle**

OUR TEAM FUN FACTOR RUBRIC

Standard	1	2	3	4	5
Speed					
Flight					
Height					
Spin					
Fun Spaces					

Name: _____   Team Name: _____

FUN FACTOR SURVEY

Next your team will use the rubric your team created to determine the fun factor of pictures of playgrounds your teacher will give you. First, fill in the blanks in each rubric below to match your team's Fun Factor Rubric. Look at the first playground image your teacher gives you. With your team, decide which word best describes the playground and circle it. Write the score for that description in the "score" column. After you have rated each standard, add up the scores to find the total fun factor for that playground. Do this for each playground image your teacher gives you.

USE FUN FACTOR TO SCORE SWING SET #1

Standard	1	2	3	4	5	Score
Speed						
Flight						
Height						
Spin						
Fun Spaces						
				Total Score/Fun Factor		

Does this swing set meet your high standards for fun?

If you could make changes to this playground, what changes would you make to improve the fun factor score?

Name: _____   Team Name: _____

STUDENT HANDOUT, PAGE 5

FUN FACTOR SURVEY

USE FUN FACTOR TO SCORE SWING SET #2

Standard	1	2	3	4	5	Score
Speed						
Flight						
Height						
Spin						
Fun Spaces						
				Total Score/Fun Factor		

Does this swing set meet your high standards for fun?

If you could make changes to this playground, what changes would you make to improve the fun factor score?

Name: _____   Team Name: _____

STUDENT HANDOUT, PAGE 6

FUN FACTOR SURVEY

USE FUN FACTOR TO SCORE SWING SET #3

Standard	1	2	3	4	5	Score
Speed						
Flight						
Height						
Spin						
Fun Spaces						
					Total Score/Fun Factor	

Does this swing set meet your high standards for fun?

If you could make changes to this playground, what changes would you make to improve the fun factor score?

Name: _____   Team Name: _____

STUDENT HANDOUT, PAGE 7

FUN FACTOR SURVEY

Take a trip out to the school playground and score your own playground swing set.

USE FUN FACTOR TO SCORE THE SCHOOL PLAYGROUND

Standard	1	2	3	4	5	Score
Speed						
Flight						
Height						
Spin						
Fun Spaces						
				Total Score/Fun Factor		

Does this swing set meet your high standards for fun?

If you could make changes to this playground, what changes would you make to improve the fun factor score?

Swing Set Makeover Lesson Plans

Name: _____   Team Name: _____

STUDENT HANDOUT

FUN FACTOR BAR GRAPH TEMPLATE

Create a bar graph of your swing set fun factor scores. Fill in the names of each swing set (swing set #1, swing set #2, swing set #3, school playground) on the horizontal axis. Fill in the possible scores on the vertical axis (the highest value should be the highest possible fun factor score for each swing set) Create a bar for each swing set's total fun factor scores.

Name: _____   Team Name: _____

STUDENT HANDOUT

FORCES PUSH BACK

SAFETY NOTES

1. All laboratory occupants must wear safety glasses or goggles during all phases of this inquiry activity.
2. Use caution when working out in the field as there can be several trip/fall or slip/fall hazards (e.g., sports equipment, uneven ground, holes) that can cause physical injuries.
3. Make sure that there are no fragile materials in the area where activities are taking place.
4. Have an appropriate level of adult supervision to ensure safe behaviors during activities.
5. Make sure all materials are put away after completing the activity.
6. Wash hands with soap and water after completing this activity.

Fill out the table. Write the sport in each gray box and circle whether a push, pull, or both is being used. Provide examples of inertia and motion. Circle the outside force(s) you identified (gravity or friction).

Sport:			Push	Pull	Both
An example of inertia is _____					
An example of motion is _____					
Outside force(s) at work:	Gravity	Friction			
Sport:			Push	Pull	Both
An example of inertia is _____					
An example of motion is _____					
Outside force(s) at work:	Gravity	Friction			
Sport:			Push	Pull	Both
An example of inertia is _____					
An example of motion is _____					
Outside force(s) at work:	Gravity	Friction			

Continued

Sport:			Push	Pull	Both
An example of inertia is _____					
An example of motion is _____					
Outside force(s) at work:	Gravity	Friction			
Sport:			Push	Pull	Both
An example of inertia is _____					
An example of motion is _____					
Outside force(s) at work:	Gravity	Friction			
Sport:			Push	Pull	Both
An example of inertia is _____					
An example of motion is _____					
Outside force(s) at work:	Gravity	Friction			
Sport:			Push	Pull	Both
An example of inertia is _____					
An example of motion is _____					
Outside force(s) at work:	Gravity	Friction			
Sport:			Push	Pull	Both
An example of inertia is _____					
An example of motion is _____					
Outside force(s) at work:	Gravity	Friction			

What roles do inertia, gravity, and friction play in keeping an object in motion?

Name: _____   Team Name: _____

Park we will present: _____

STUDENT HANDOUT

PARK EXHIBITION SHOWCASE

You will share your discoveries about a local park with your classmates. Be sure to research, show pictures, and share artifacts that will engage your audience and cause them to want to visit your park today.

Consider what they may want to know and the best way to present the park. For example, for your classmates to visit the park, they need to know where it is located. You can show the address on your display or use a map and mark its location. How you do it will be up to you. For example, you can present facts using a trifold brochure, a poster, a one-page flyer, or note cards taped onto cardboard that will allow visitors to flip the cards to get information. Be creative!

BEGIN WITH YOUR RESEARCH
What are your research questions? Information you may want to research includes the following:

- How this park got its name
- Information about the park's location
- Natural features in the park that make it unique
- The kind of trees, flowers, and wildlife at this park
- Special landmarks or features that make this park special
- Services provided by park staff or volunteers
- Park activities that can be done at different seasons of the year
- Any fun or interesting stories associated with this park

COLLECT ARTIFACTS
Use pictures, objects, flyers, and photographs to help tell the viewers about the park. Use labels to direct your visitors to view the materials in the order you think will be most helpful.

- What items are interesting or cool to look at and touch?
- What artifacts help share an interesting fact?
- What artifacts will make the presentation better?
- Can I use an artifact to tell a story about my trip to the park?

PARTICIPATION REQUIRED
This is a group project. How are you going to show the teacher and your classmates that everyone worked together to create your exhibit? What can you add to your exhibit to invite others to participate? Consider adding a game, quiz facts, or unusual artifacts so others can engage in new experiences. Use your imagination!

STEM Research Notebook Rubric	
Name: _____	
10	**OUTSTANDING** • The writing goes beyond the basic requirements and shows in-depth understanding of concepts. • The work shows in-depth reflection throughout the learning process. • The notebook has all the components expected, including dates and labels on each page. • All pages are numbered properly, with odd numbers on the right and even numbers on the left. • Work is correctly organized with all criteria. • Use of color and labeled diagrams enhance understanding. • The notebook is very neat.
9	**GREAT** • The writing follows the basic requirements, shows understanding of concepts, but does not go beyond. • The work shows in-depth reflection. • The notebook has all the components expected, including dates and labels on each page. • All pages are numbered properly, with odd numbers on the right and even numbers on the left. • Work is correctly organized with all criteria. • The notebook has color and uses labeled diagrams. • A 9 looks much like a 10 but it lacks the perfection of a 10.
8	**GOOD** • The written work shows a basic understanding of concepts. • The work is an honest reflection, but it is limited. • The notebook has about 90% of the components expected, with dates and labels. • All pages are numbered properly, with odd numbers on the right and even numbers on the left. • Work is correctly organized. • The notebook has some color and diagrams and a few labels. • Some requirements are met, but the notebook does not meet all criteria.

Continued

STEM Research Notebook Rubric (*continued*)

7	OK
	• The written work shows a limited understanding of concepts.
	• The work shows limited reflection overall.
	• The notebook has about 80% of the components, with dates and labels.
	• Most pages are numbered.
	• Work is fairly organized, just so-so.
	• The notebook has very little color and few diagrams.
	• Notebook requirements are rarely met.
6	NEEDS IMPROVEMENT
	• The written work shows misconceptions and a lack of understanding.
	• The work includes little-to-no reflection.
	• The pages in the notebook are unfinished.
	• There are incomplete dates and labels.
	• Most pages are not numbered or are numbered inconsistently.
	• The notebook is unorganized and one or two pages are blank.
5	INCOMPLETE
	• Many pages are blank or include the class template only.

SCORE: _____

COMMENTS:

Bar Graph Rubric			
Name: _____			
Criteria	Needs Improvement	Approaching Standard	Meets Standard
THE GRAPH IS NEAT. • Lines are straight. • Labels can be read easily. • Coloring is neat.	1	2	3
THE GRAPH IS COMPLETE. • Graph has a title. • Graph is the right size. • A legend or color-coding is used.	1	2	3
HORIZONTAL AXIS IS COMPLETE. • The axis is labeled. • The correct labels are used. • The intervals are equal.	1	2	3
VERTICAL AXIS IS COMPLETE. • The axis is labeled. • The units are correct. • The intervals are equal.	1	2	3
THE GRAPH IS ACCURATE. • The points/bars stop at the right place. • The lines/bars are the same size/width. • The scale used to show units makes sense.	1	2	3
TOTAL SCORE: _____ **COMMENTS:**			

Team Name: _____

Park Presentation Rubric (30 points possible)

Criteria	Needs Improvement (1–2 points)	Approaching Standard (3–4 points)	Meets or Exceeds Standard (5–6 points)	Team Score
INFORMATION	• Few details about the park are given. • Nothing seems to be special about this park.	• Some details about the park (location, size, types of plants and animals) are included. • Ways this park is special are provided.	• Many details about the park (location, size, types of plants and animals) are included. • Ways this park is special (including programs and events) are provided.	
PRESENTATION	• Information is random and no strategy is present. • No introduction is used. • No conclusion is provided. • Much of the time is wasted due to lack of preparation. • The presentation is very short or long.	• Focus on information wanders. • An introduction is used. • A conclusion summary is provided. • Some parts are hard to understand. • The time is used satisfactorily. • The presentation is slightly less than 5 minutes or longer than 7 minutes.	• Information is focused and clear. • The introduction gets the audience's attention. • The conclusion summarizes points. • Presentation is easy to understand. • The time is used well and it is obvious the team has practiced. • Presentation is between 5 and 7 minutes long.	
PARTICIPATION	• Only one or two team members share. • Presenters are difficult to understand. • No science or key terms are used.	• Not all team members share. • Volume may be too low or presenters speak too fast. • Presenters use science words and identify some scientific observations.	• All team members share. • Presenters are easy to understand. • Presenters use science words and key park terms while making scientific observations about the park.	
VISUALS	• Visual aids are poorly presented or distract from the presentation.	• Visual aids are used but do not add to the presentation.	• Visual aids or media used add much to the presentation.	
QUESTIONS	• Team fails to respond to questions from audience or responds inappropriately.	• Team responds appropriately to audience questions. • Responses may be brief, incomplete, or unclear.	• Team respectfully seeks clarification of questions when required. • Team answers questions in detail.	

TOTAL SCORE: _____

COMMENTS:

Lesson Plan 2: Slippery Slide Design

In this lesson, students investigate Newton's laws of motion and apply their learning to make a more efficient ramp on which a car will travel. This learning activity is directly connected to the final challenge in which students build a model of the slide they design during this lesson. In mathematics, students record and graph data. They also note the geometric shapes on a swing set, taking measurements as an introduction to area measurements. In ELA, they write an opinion paper to prepare for the final blog activity connected to the Swing Set Makeover Design Challenge. In social studies class, students finish their park research projects and build their exhibits.

ESSENTIAL QUESTIONS

- How does the shape and incline of a ramp affect an object's motion?

- What materials work best to minimize friction on a ramp?

- What improves the fun factor of a slide design?

- How can the shape of the slide be changed to effectively use balanced and unbalanced forces?

- What are some of the safety concerns when developing a slide design?

ESTABLISHED GOALS AND OBJECTIVES

At the conclusion of this lesson, students will be able to do the following:

- Identify geometric shapes in playground equipment

- Design and implement an investigation around a testable question

- Use data analysis as evidence to answer a question under investigation

- Relate the angle and length of a ramp to increased and decreased speed

- Calculate the area of a surface

- Explain that friction causes work (energy) to be wasted when objects go down the slide

- Compare and contrast materials and angles to make decisions for a new slide design

- Critique student slide designs of classmates to provide constructive feedback

- Formulate opinion statements and provide supporting reasons for opinions

TIME REQUIRED

- 6 days (approximately 45 minutes each day; see Tables 3.6 and 3.7, p. 41)

MATERIALS

Handouts for Lesson 2

- Ramp Investigation #1 and #2 (1 of each per student)

- Engineering Design Process visual (1 per team)

- Slide Makeover Plan Book (1 per student)

- Geometry Scavenger Hunt (1 per team)

- Writing the OREO Way (1 per student)

- Park Exhibition Showcase (1 per student; see Lesson 1, p. 87)

Rubrics for Lesson 2

- Collaboration Rubric

- Ramp Investigations Rubric

- Slide Design Sketch Rubric

Necessary Materials for Lesson 2

- STEM Research Notebooks (1 per student)

- *Roll, Slope, and Slide: A Book About Ramps* by Michael Dahl (Capstone, 2006)

- Access to the internet for showing videos (optional) and student research

- KWL chart from previous lesson

- Optional: Bill Nye DVDs may offer additional support to help students understand concepts in this module. For example, *Bill Nye the Science Guy: Friction Classroom Edition* and *Bill Nye the Science Guy: Gravity Classroom Edition.* These can be found by searching online.

- 1 magnifying lens (per team)

- 1 marble (per team)

- 1 drinking straw (per student)

Additional Materials for Ramp Investigations

- 1 ramp per team; ramps should be 4–6 inches wide and 24–30 inches long. Ramps can be made of balsa wood or rigid plastic strips.

- Classroom items, possibly textbooks, to build up ramps (per team)

- 1 sheet (per team) of wax paper, aluminum foil, sandpaper, and felt in sizes sufficient to cover ramps

- Classroom supply of plastic wrap, wrapping paper, shelf liner, or fabric in sizes sufficient to cover ramps

- 1 block at least ½ inch smaller than the width of the ramps, tightly wrapped in a smooth fabric (e.g., unfeatured polyester or cotton fabric) (per team)

- 1 toy car or truck per team (small enough to fit on the ramps)

- 1 roll of masking tape (per team)

- 1 tape measure or meter stick (per team)

- 1 pair of scissors (per team)

- Safety glasses or goggles (1 per student)

Additional Materials for Geometry Scavenger Hunt

- Paper for anchor chart (per student)

- 1–2 clipboards (per team)

- 1–2 pencils (per team)

- 1 tape measure (per team)

- 1 ruler (per team)

- 1 protractor (per team)

- 1 calculator (per team)

Additional Materials for Park Exhibition Showcase

- Internet access for research

- 1 poster board or poster-size paper (per team)

- 1 glue stick (per team)

- 2–3 markers (per team)

Optional Literature Connections

- *Roll, Slope, and Slide: A Book About Ramps* by Michael Dahl (Capstone, 2006)

- *Roller Coaster* by Marla Frazee (HMH Books for Young Readers, 2006)

- *I Want a Dog: My Opinion Essay* by Darcy Pattison (Mims House, 2015)

SAFETY NOTES

1. All laboratory occupants must wear safety glasses or goggles during all phases of this inquiry activity.

2. Use caution when working out in the field as there can be several trip/fall or slip/fall hazards (e.g., sports equipment, uneven ground, holes) that can cause physical injuries.

3. Make sure there are no fragile materials in the area where activities are taking place.

4. Have an appropriate level of adult supervision to ensure safe behaviors during activities.

5. Use caution when working with sharps (e.g., scissors, sticks) to avoid cutting or puncturing skin or eyes.

6. Make sure all materials are put away after completing the activity.

7. Wash hands with soap and water after completing this activity.

CONTENT STANDARDS AND KEY VOCABULARY

Table 4.7 (p. 96) lists the content standards from the *NGSS, CCSS,* and the Framework for 21st Century Learning that this lesson addresses, and Table 4.8 (p. 100) presents the key vocabulary. Vocabulary terms are provided for both teacher and student use. Teachers may choose to introduce some or all of the terms to students.

Table 4.7. Content Standards Addressed in STEM Road Map Module
Lesson 2

NEXT GENERATION SCIENCE STANDARDS

PERFORMANCE EXPECTATIONS

- 3-PS2-1. Plan and conduct an investigation to provide evidence of the effects of balanced and unbalanced forces on the motion of an object.

- 3-PS2-2. Make observations and/or measurements of an object's motion to provide evidence that a pattern can be used to predict future motion.

- 3-5-ETS1-1. Define a simple design problem reflecting a need or a want that includes specified criteria for success and constraints on materials, time, or cost.

- 3-5-ETS1-2. Generate and compare multiple possible solutions to a problem based on how well each is likely to meet the criteria and constraints of the problem.

SCIENCE AND ENGINEERING PRACTICES

Asking Questions and Defining Problems

- Identify scientific (testable) and non-scientific (non-testable) questions.

- Ask questions that can be investigated and predict reasonable outcomes based on patterns such as cause and effect relationships.

- Define a simple design problem that can be solved through the development of an object, tool, process, or system and includes several criteria for success and constraints on materials, time, or cost.

Developing and Using Models

- Collaboratively develop and/or revise a model based on evidence that shows the relationships among variables for frequent and regular occurring events.

- Develop a model using an analogy, example, or abstract representation to describe a scientific principle or design solution.

- Develop a diagram or simple physical prototype to convey a proposed object, tool, or process.

- Use a model to test cause and effect relationships or interactions concerning the functioning of a natural or designed system.

Planning and Carrying Out Investigations

- Plan and conduct an investigation collaboratively to produce data to serve as the basis for evidence, using fair tests in which variables are controlled and the number of trials considered.

- Make observations and/or measurements to produce data to serve as the basis for evidence for an explanation of a phenomenon or test a design solution.

- Make predictions about what would happen if a variable changes.

Continued

Table 4.7. (*continued*)

Analyzing and Interpreting Data

- Represent data in tables and/or various graphical displays (bar graphs, pictographs, and/or pie charts) to reveal patterns that indicate relationships.

- Analyze and interpret data to make sense of phenomena, using logical reasoning, mathematics, and/or computation.

- Compare and contrast data collected by different groups in order to discuss similarities and differences in their findings.

Using Mathematics and Computational Thinking

- Decide if qualitative or quantitative data are best to determine whether a proposed object or tool meets criteria for success.

- Organize simple data sets to reveal patterns that suggest relationships.

Constructing Explanations and Designing Solutions

- Use evidence (e.g., measurements, observations, patterns) to construct or support an explanation or design a solution to a problem.

- Identify the evidence that supports particular points in an explanation.

- Apply scientific ideas to solve design problems.

- Generate and compare multiple solutions to a problem based on how well they meet the criteria and constraints of the design solution.

Engaging in Argument From Evidence

- Compare and refine arguments based on an evaluation of the evidence presented.

- Distinguish among facts, reasoned judgment based on research findings, and speculation in an explanation.

- Respectfully provide and receive critiques from peers about a proposed procedure, explanation, or model by citing relevant evidence and posing specific questions.

- Construct and/or support an argument with evidence, data, and/or a model.

- Use data to evaluate claims about cause and effect.

- Make a claim about the merit of a solution to a problem by citing relevant evidence about how it meets the criteria and constraints of the problem.

DISCIPLINARY CORE IDEAS

PS2.A: Forces and Motion

- Each force acts on one particular object and has both strength and a direction. An object at rest typically has multiple forces acting on it, but they add to give zero net force on the object. Forces that do not sum to zero can cause changes in the object's speed or direction of motion.

- The patterns of an object's motion in various situations can be observed and measured; when that past motion exhibits a regular pattern, future motion can be predicted from it.

Continued

Table 4.7. *(continued)*

PS2.B: Types of Interactions

- Objects in contact exert forces on each other.

CROSSCUTTING CONCEPTS

Cause and Effect

- Cause and effect relationships are routinely identified, tested, and used to explain change.

Patterns

- Patterns of change can be used to make predictions.
- Patterns can be used as evidence to support an explanation.

Structure and Function

- Different materials have different substructures, which can sometimes be observed.
- Substructures have shapes and parts that serve functions.

Influence of Science, Engineering, and Technology on Society and the Natural World

- People's needs and wants change over time, as do their demands for new and improved technologies.
- Engineers improve existing technologies or develop new ones to increase their benefits, to decrease known risks, and to meet societal demands.

COMMON CORE STATE STANDARDS FOR MATHEMATICS

MATHEMATICAL PRACTICES

- MP1. Make sense of problems and persevere in solving them.
- MP2. Reason abstractly and quantitatively.
- MP4. Model with mathematics.
- MP5. Use appropriate tools strategically.
- MP7. Look for and make use of structure.

MATHEMATICAL CONTENT

- 3.MD.A.2. Measure and estimate liquid volumes and masses of objects using standard units of grams (g), kilograms (kg), and liters (l). Add, subtract, multiply, or divide to solve one-step word problems involving masses or volumes that are given in the same units, e.g., by using drawings (such as a beaker with a measurement scale) to represent the problem.
- 3.MD.B.4. Generate measurement data by measuring lengths using rulers marked with halves and fourths of an inch. Show the data by making a line plot, where the horizontal scale is marked off in appropriate units—whole numbers, halves, or quarters.
- 3.MD.C.5. Recognize area as an attribute of plane figures and understand concepts of area measurement.

Continued

Table 4.7. (*continued*)

COMMON CORE STATE STANDARDS FOR ENGLISH LANGUAGE ARTS

READING STANDARDS

- RI.3.5. Use text features and search tools (e.g., key words, sidebars, hyperlinks) to locate information relevant to a given topic efficiently.

- RI.3.7. Use information gained from illustrations (e.g., maps, photographs) and the words in a text to demonstrate understanding of the text (e.g., where, when, why, and how key events occur).

- RI.3.10. By the end of the year, read and comprehend informational texts, including history/social studies, science, and technical texts, at the high end of the grades 2–3 text complexity band independently and proficiently.

WRITING STANDARDS

- W.3.1. Write opinion pieces on topics or texts, supporting a point of view with reasons.

- W.3.1.A. Introduce the topic or text they are writing about, state an opinion, and create an organizational structure that lists reasons.

- W.3.1.B. Provide reasons that support the opinion.

- W.3.1.C. Use linking words and phrases (e.g., *because, therefore, since, for example*) to connect opinion and reasons.

- W.3.1.D. Provide a concluding statement or section.

- W.3.2. Write informative/explanatory texts to examine a topic and covey ideas and information clearly.

- W.3.2.B. Develop the topic with facts, definitions, and details.

- W.3.3. Write narratives to develop real or imagined experiences or events using effective technique, descriptive details, and clear event sequences.

- W.3.7. Conduct short research projects that build knowledge about a topic.

- W.3.8. Recall information from experiences or gather information from print and digital sources; take brief notes on sources and sort evidence into provided categories.

SPEAKING AND LISTENING STANDARDS

- SL.3.1. Engage effectively in a range of collaborative discussions (one-on-one, in groups, and teacher-led) with diverse partners on *grade 3 topics and texts,* building on others' ideas and expressing their own clearly.

- SL.3.1.D. Explain their ideas and understanding in light of the discussion.

Continued

Table 4.7. (*continued*)

FRAMEWORK FOR 21ST CENTURY LEARNING
• Interdisciplinary Themes: Health and Safety; Environmental Literacy; Science; Mathematics
• Learning and Innovation Skills: Creativity and Innovation; Critical Thinking and Problem Solving; Communication and Collaboration
• Information, Media, and Technology Skills: Information Literacy; Media Literacy
• Life and Career Skills: Flexibility and Adaptability; Initiative and Self-Direction; Social and Cross-Cultural Skills; Productivity and Accountability; Leadership and Responsibility

Table 4.8. Key Vocabulary for Lesson 2

Key Vocabulary	Definition
action force	the force that is acting on an object in one direction
acute angle	an angle in which the space between the angle's rays is between 0 and 90 degrees
anchor charts	a classroom tool usually cocreated by the teacher and students that visually displays key concepts and vocabulary
annotated sketch	a drawing that displays the creator's thinking by including labels and arrows that lead to explanations of details or other notes
area	the amount of space found inside of a shape
constant velocity	when an object travels the same distance every second
distance	the measure of how far an object has traveled
driving force	the energy behind something in motion
edge	the outer limit of a space or the place where something stops
energy	the ability of a body or system to do work or produce a change; a power exerted with force
equal	when things are the same in number, amount, degree, or quality
geometry	an area of mathematics that deals with points, lines, angles, and surfaces
incline	a slanted surface that goes up
inclined plane	a simple machine with a slanted surface that connects a lower level area to a higher level area

Continued

Table 4.8. (*continued*)

Key Vocabulary	Definition
iteration	revisiting previous steps in a problem-solving or design process to improve results
mass	a measure of how much matter is in an object; on Earth, mass equals weight
obtuse angle	an angle in which the space between the angle's rays is between 90 and 180 degrees
opposite	to be in a position on the other side or set to work against something
perimeter	the total distance around the edge of a two-dimensional shape
polygon	a two-dimensional closed shape with three or more sides made of straight lines
quadrilateral	a two-dimensional shape with four sides
ramp	a slope for joining two levels
reaction force	the force that is acting on an object in the opposite direction of the action force
responsibility	a character trait that includes acting in trustworthy and thoughtful ways and accepting the consequences of one's actions
right angle	an angle in which the space between the angle's rays is 90 degrees
stationary	something that is not moving
technology	the use of science to invent useful things or solve problems
weight	the amount of the downward force on a body

TEACHER BACKGROUND INFORMATION
Balanced and Unbalanced Forces

Students will continue to explore balanced and unbalanced forces in this lesson. Balanced forces occur when two forces are pushing in opposite directions, resulting in no movement. When forces have different strengths (unbalanced), then there is movement in the direction of the greater force.

Examples of balanced and unbalanced forces include the following:

Balanced Forces	Unbalanced Forces
• Two people hugging	• Arm wrestling—one person stronger than the other
• Scale with two items of equal weight	• Soccer ball flying through the air after being kicked
• Book sitting on a table—equal force pushing up and down to make it sit still	• Seesaw—one person kicking off and creating more force than the weight of the person on the other end, which creates up and down motion
• Two dogs pulling on a rope when they are equally matched	
• A rock setting in your hand—force holding it in the air to match gravity's pull	• A person standing on a skateboard being pushed forward by another person
	• Pushing a door closed

You may wish to use the "Balanced and Unbalanced Forces" video from the Fuse School (*www.youtube.com/watch?v=YyJSlclbd-s*) to display some of the forces of nature that are balanced and unbalanced and introduce students to new science vocabulary. The video also provides examples of how to illustrate balanced and unbalanced forces using arrows.

Engineering Design Process

The engineering design process (EDP) is a series of steps that engineers follow when they are trying to solve a problem. The solution often involves designing a product (such as a machine or a computer code) that needs to meet certain criteria or accomplish a particular task. A graphic representation of the EDP is provided in Chapter 2 and as a handout at the end of this lesson; it may be useful to post it in a prominent place in your classroom for student reference throughout the module. Be prepared to review the steps of the EDP with students, and emphasize that the process is not a linear one—at any point in the process they may need to return to a previous step. See Chapter 2's Engineering Design Process section (p. 9) for more information on the EDP steps.

Engineers do not always follow the EDP steps in order. It is very common to design something, test it, find a problem, and then go back to an earlier step to make a modification or change the design entirely. This way of working is called *iteration*.

Playgrounds and Technology

As students progress through the module, they will begin to understand that swing sets incorporate a number of technological innovations. For example, playgrounds incorporate innovative materials including resins or plastics, such as materials covering metal bars, and materials incorporating recycled rubber. These innovations enhance safety, reduce maintenance, and increase the usable life of equipment and may introduce new features to enhance user experience (e.g., zip lines). Students should understand that technology is the use of science to create things or to solve problems. If students have difficulty associating swing sets and other playground equipment with technology, viewing images of older playground equipment (e.g., equipment from the 1950s) beside images of today's playground equipment can highlight the technological innovations. Therefore, you may wish to have images of older playground equipment on hand to compare with the modern swing set images attached at the end of Lesson 1.

Because student teams will be designing slides for their Swing Set Makeover Design Challenge in this lesson, you may wish to share some images or videos of various types of slides to inspire students. The following are examples of videos that can help inform student design ideas:

Water Slides

- *www.youtube.com/watch?v=hhk2uqz5umA*

- *www.youtube.com/watch?v=JA8iBapiQRo*

Straight Slides

- *www.youtube.com/watch?v=r56AHblQvlo*

- *www.youtube.com/watch?v=jmtWo9nRLEE*

- *www.youtube.com/watch?v=iio5zXVgiUI*

- *www.youtube.com/watch?v=eRh4DVbMcNY*

Curved and Spiral Slides

- *www.youtube.com/watch?v=U-q_Qvn0iTc*

- *www.youtube.com/watch?v=KeLpPljQCK4*

- *www.youtube.com/watch?v=15oG0yU8kcU*

Tunnel Slides

- *www.youtube.com/watch?v=KBP7LMfhB6k*

- *www.youtube.com/watch?v=jM3WqbY3glY*

Mass and Weight

Understanding mass and weight can be challenging for third graders. If you wish to introduce these concepts to students, you can begin to address potential misunderstanding by listing the weights of famous athletes on Earth. Discuss the difference between

weight and mass, and then visit the website *www.exploratorium.edu/ronh/weight* to discover the athletes' weights on other planets. Students can create graphs showing the difference and develop a hypothesis about why the athletes' masses are the same but their weights are not the same on all planets.

Writing

Third-grade writing standards focus on the writing process as the primary tool to help children become independent writers. In grade 3, students can use the following process:

- *Prewriting Phase.* Students consider the purpose and intended audience, and then make a plan for writing that includes the topic. They use prewriting techniques to generate ideas such as drawing and listing key thoughts.

- *Drafting Phase.* Students write several drafts before producing a final product. In drafting, students strive to develop the main idea with supporting details while organizing information into a logical sequence using words to reveal timing and cause-and-effect transitions. In addition, student writing should reflect the topic, audience, and purpose.

- *Revision Phase.* Students revise their writing to improve consistency, organization, voice (formal or informal), and effectiveness. Students edit to use more precise words, clarify details, incorporate variety, and implement effective literary devices to create interest.

- *Editing Phase.* Students edit and correct the draft for appropriate grammar, spelling, punctuation, capitalization, and other features of polished writing.

- *Publishing the Final Product.* Students produce, illustrate, and share a variety of compositions, using the appropriate technology to compose and publish their work.

Writing to Share Opinions

Some teachers have adopted a strategy for writing opinion papers called the "OREO strategy." Genia Connell wrote a blog for Scholastic on Opinion Writing for third graders that uses this strategy (*www.scholastic.com/teachers/blog-posts/genia-connell/graphic-organizers-opinion-writing*). Included in her comments are the following pointers:

- Introduce the language of opinion writing, including introductory phrases, transitions, and opinion clues.

- Introduce easy-to-read opinion pieces (she suggests several, including "Should the Penny Be Retired?" and "Should Students Learn a Foreign Language?")

- Model, model, model!

Reviews are one of the most common types of content on the internet and represent an example of opinion writing. When buying a product, many people read the reviews on multiple websites including Amazon. The most common reason people look for product reviews include the following:

- Learn about the pros and cons of a given product.

- Find out the quality of the product and if it is easy to use.

- Discover if there is a cheaper, yet compatible, substitute.

- Learn about other peoples' experiences with the item.

Reviews are a public way to share opinions and provide feedback to manufacturers.

Teaching Strategies

Groups to Support Learning. You may wish to maintain the student groupings from Lesson 1 or regroup students according to one of the strategies described in Chapter 3 (see the Strategies for Differentiating Instruction Within This Module section, p. 33). Students will begin to use the EDP during the slide makeover portion of the overall Swing Set Makeover Design Challenge and will continue to use the EDP throughout the module as they brainstorm and sketch each component of the swing set model for their final challenge. You should therefore group students in Lesson 2 into the design teams that they will maintain throughout the remainder of the module.

Anchor Charts and Visual Representation of Key Concepts. Communication is an important process in science and mathematics. To effectively communicate, students need to represent academic thinking concretely, pictorially, symbolically, physically, verbally, and in writing. Because there are so many ways to connect, teachers need to help students learn these different ways to communicate. Anchor charts are a tool that can be used to "anchor" student thinking while providing an opportunity to model thinking and build vocabulary.

In this lesson, anchor charts are used to review the concepts of perimeter and area. This can be a valuable scaffolding project since this provides a nonthreatening way for students to demonstrate their understanding as well as a way for students to teach one another as they work on their charts. Anchor charts in mathematics provide a visual reference to support student thinking, reasoning, and problem solving. An effective anchor chart

- is focused on a single concept,

- reflects recent lesson content,

- helps students remember a skill or a strategy,

- supports science or mathematics language,

- is organized and accurate, and

- is organized with the students' input.

Student Research. You may wish to scaffold group research projects by providing ways for students to divide work among themselves. In the Park Exhibition Showcase activity, for example, you may wish to invite each group to create a list of ideas for research topics on chart paper (e.g., animal life, plant life, activities available, playground equipment). Each team member then chooses a topic and writes a research question to investigate, such as the following: What animals live at this park? What can we do at this park? Does this park have a swing set? What makes this park special?

COMMON MISCONCEPTIONS

Students will have various types of prior knowledge about the concepts introduced in this lesson. Table 4.9 outlines some common misconceptions students may have concerning these concepts. Because of the breadth of students' experiences, it is not possible to anticipate every misconception that students may bring as they approach this lesson. Incorrect or inaccurate prior understanding of concepts can influence student learning in the future, however, so it is important to be alert to misconceptions such as those presented in the table.

Table 4.9. Common Misconceptions About the Concepts in Lesson 2

Topic	Student Misconception	Explanation
Engineers and the engineering design process (EDP)	Engineers are people who drive trains.	An engineer is someone who uses science, technology, and mathematics to build machines, products, and structures that meet people's needs.
Mass and weight	Mass and weight are the same thing.	An object's weight is how much gravity is pulling on it while mass is the amount of matter in an object. An object on Earth and in space might have the same mass but different weights since as an object moves farther from Earth it does not experience the gravitational force of the Earth as strongly.
Friction	Friction always opposes (works against) motion.	Although friction is a force that opposes motion, it can also play a crucial role in allowing motion. For example, walking or riding a bike would not be possible without friction between a person's feet and a sidewalk or between a bike's wheels and the road surface.

PREPARATION FOR LESSON 2

Review the Teacher Background Information provided (p. 101), assemble materials for the lesson and preview videos included within the Learning Components section.

In Ramp Investigation #1: Ramp Height, students will roll toy cars down ramps and measure the distances they travel. Ensure that you have access to an area with adequate open space for the cars to travel freely without encountering obstacles.

Have on hand a few images of older playground equipment (e.g., images from the 1950s, 1960s, 1970s, and 1980s) to highlight the technological innovations in playgrounds. Comparing these images with the swing set images attached at the end of Lesson 1 may be useful to help students understand that swing sets incorporate technological innovations to enhance safety, reduce maintenance, increase the usable life of playground equipment, and introduce new features to enhance user experience (e.g., zip lines).

Read through the Geometry Scavenger Hunt handout (p. 133) to make sure the questions can all be answered as written. Some of the shapes listed on the Geometry Scavenger Hunt handout may not be found on your school playground and modifications may therefore be necessary to customize the activity to the playground equipment available at your school. The Geometry Scavenger Hunt does not have an answer key. Your modifications will need to be made before completing the answers for a key.

Student teams will use reviews of swing sets to create an opinion paper. Locate several reviews of swing sets or playground equipment and make copies for each student team, or be prepared to have teams read these reviews online.

For the Park Exhibition Showcase activity, make arrangements with the school librarian to set up a display in the library with all or the best of the park exhibits.

LEARNING COMPONENTS

Introductory Activity/Engagement

Connection to the Challenge: Begin each day of this lesson by directing students' attention to the driving question for the module and challenge: How can I use what I know about force and motion to create a plan and build a model of a swing set that is both fun and safe? Discuss what students learned in Lesson 1 that will be helpful in addressing the challenge. Have students record their ideas in their STEM Research Notebooks. Next, review the driving question for Lesson 2 (see next page), ask students to share their thoughts, and have a class discussion to lead into the concepts for Lesson 2. Ask the students the following:

- Is it true that everything that moves will eventually slow down and stop? If not, give an example of an object that keeps moving and never stops. (*Planets in space move in an orbit and never stop.*)

- Can the shape of a ramp affect the rate at which an object travels on this ramp? Explain your reasoning.

Driving Question for Lesson 2: How can I use what I learned about friction and balanced and unbalanced forces when modifying an inclined plane to create a plan and build a model for a slide that is both fun and safe?

Science Class: Review the concept that there is almost always more than one force acting on an object (e.g., when you drop a ball, there is gravity and wind resistance or friction) by watching the following video of a marble race: *www.youtube.com/watch?v=UaMr6bFruF0.*

Students may recall from the Forces Push Back activity in Lesson 1 that some forces act when they come into direct contact with an object, such as when a box is moved. Other forces don't need to come into contact with an object to act, such as a jumper who leaps from a plane, since gravity pulls him or her toward the ground without any contact at all. Have students draw a T-chart with "Contact" and "No Contact" headings. Have them provide one or two examples of each in the appropriate columns.

Review the concept of friction (a force that occurs when surfaces rub together) with students. Different surfaces can apply more or less friction, depending on how smooth or rough they are. Distribute a variety of materials with different surfaces to each design team (e.g., rough sandpaper, cardboard, fabric, wax paper, aluminum foil, plastic wrap). Have the teams sort the materials into two groups: those with smooth surfaces and those with rough surfaces. Lead students to understand that different surfaces can apply more or less friction, depending on how smooth or rough they are. Have students use magnifying lenses to look at the surfaces they sorted and then other surfaces around the room.

Ask students to predict which surfaces would create the most friction to an object moving on it and which would create the least friction. Explain to students that friction is necessary to make some things work, like riding a bike or walking. Hold a class discussion about what would happen to cars if a road provided no friction, or what would happen to the students if the floor provided no friction. You may wish to use an example of walking or riding a bike on ice as an example of a low-friction situation.

Have students work in their design teams to make lists of things that increase friction (e.g., treads on gym shoes, sand on roads) and things that decrease friction (e.g., adding oil to gears, smoothing out the ice before a hockey game), reminding students that objects that are moving will keep moving unless a force acts on them.

Engineers use what they know about moving objects and forces to design safer cars, buildings, and even swing sets. Explain to students that engineers need to understand different types of forces as they design items. Show the "Balanced and Unbalanced Forces" video from the Fuse School (*www.youtube.com/watch?v=YyJSlcIbd-s*) to demonstrate some of the forces of nature that are balanced and unbalanced. The video serves as an introduction to new science vocabulary terms that are illustrated as each new term is introduced. You may wish to have students create a Venn diagram to demonstrate their understanding of balanced and unbalanced forces.

After watching the video, discuss the meaning of the word *balanced* and contrast this with the meaning of the word *unbalanced* with the class. Associate these terms with forces. Have students draw a cartoon in their STEM Research Notebooks showing how balanced and unbalanced forces interact. One example might be a picture of a canoe with a person in it and a dog swimming beside the canoe and a second picture of the same canoe with the dog riding inside the boat and the boat sinking lower in the water. Have the students share their pictures with their design teams.

Review the concepts of balanced and unbalanced forces. Lead students to describe balanced forces as forces acting on an object so that the object's speed and direction do not change. This can mean that an object moves at a constant speed and direction or an object remains still. Unbalanced forces are forces acting on an object so that the object's speed or direction changes.

Next, have student teams investigate balanced and unbalanced forces using a marble and straws. Distribute one marble to each design team and one straw to each student. Have the teams place a marble on a desk, and ask students to draw diagrams in their STEM Research Notebooks that show the forces acting on the marble. Students should label the diagram to show the balanced and unbalanced forces they recognize.

Then, have a student from each team use a drinking straw to blow air on the marble. The marble should move but stay on the tabletop. Instruct students to draw another diagram, again showing all of the forces acting on the marble. Have students label the diagram showing the balanced and unbalanced forces.

Finally, challenge the teams to blow on the marble so that the marble does not move. Instruct students to draw and label a diagram of forces acting on the marble and label the diagram showing the balanced and unbalanced forces.

Mathematics Connection: Introduce the role of geometric shapes in structures constructed by humans with the following questions:

- What are some shapes you see in houses and other buildings?

- Look around the classroom—what shapes do you see in the structure of the room (on the walls, doors, windows, and ceilings)?

- Why do you think these shapes are used?

- What would happen if different shapes were used—for example, if circles were used for walls instead of rectangles, or if triangles were used for doorways?

Next, show the barn door image attached at the end of the lesson (p. 132). Tell students that the builder added extra boards to make triangle shapes in both parts of the door. Ask the students the following: Why do you think these boards were added?

ELA Connection: Open the lesson by reading the book *Roll, Slope, and Slide: A Book About Ramps* by Michael Dahl, as a class. Discuss the book using the following discussion prompts:

- What is happening in the picture on the cover? Has anyone been on a skateboard before or watched someone on a skateboard? Do you see any ramps on the cover? What makes them ramps?

- As a class, create a working definition to describe a ramp.

- What are ramps used for?

- Are there examples of ramps in the classroom or on the playground? How are these ramps used to make tasks easier?

- If you could go skateboarding like the boy in the story, would you rather ride on a steep ramp or one that is less steep? Why?

Social Studies Connection: Hold a class discussion to review students' findings in their Park Exhibition Showcase research, using questions and prompts such as the following:

- The playground is a favorite part of a park for most children. Why isn't a park just one big playground?

- What kinds of structures might you discover in the park? How do they help people enjoy the park? (*pavilions, playgrounds, sports fields, statues, monuments, gardens, fountains, trails*)

Activity/Exploration

Science Class: Introduce the Ramp Investigation activities by showing students three items: a ball, a pencil, and a block. Hold a class discussion, asking student to respond to the following:

- How does each of these items represent ways that a person might travel down a ramp? (*The way the ball travels is similar to a person doing somersaults down a hill, a pencil is similar to a log rolling down a hill, and block is similar to a person sliding down a slide.*)

- Which of these would move fastest down a ramp? Have students explain their reasoning using science key terms such as *balanced* and *unbalanced forces, friction,* and *motion.* Test student predictions by demonstrating the motion of the objects down a ramp and having students make observations.

- Were the results as expected? (Allow students to answer and offer explanations for their observations. Students should agree the block is the slowest.)

- How can friction be reduced on the block? (*Friction can be reduced by putting wheels under it—like a roller coaster, or making the surface more slippery with ice, water, or snow—like a water slide.*)

- How can you decrease the speed of the fastest object? (*Make the surface rougher by putting sand on it; make the path wider so the ball bounces around more.*)

- What parts of playground equipment are ramps? (*sliding boards, walkways*)

Tell students that their swing set design for the module challenge will include a slide. Ask students to brainstorm with their design team what they could investigate to help them make the best decision for their slide design. (*Guide students to address slide incline or slope, slide surface material, and safety.*) This lesson has two separate investigations students will complete in their design teams that will allow them to explore features of slide design. The first focuses on ramp height and the second focuses on ramp surface.

Ramp Investigation #1: Ramp Height

In this investigation, students will experiment with various ramp inclines using a toy car and a block covered in smooth fabric such as an untextured polyester or cotton (to decrease friction and simulate slider riders' clothing) as their "riders." Distribute the Ramp Investigation #1 handout to each student and investigation materials to each team. Teams will test their slides by measuring how far beyond the end of the slide the "rider" travels. Remind students that they will apply their findings about the ramp slopes to their slide designs. As teams test the various slopes, they should be mindful of the safety of their rider (i.e., some inclines might be very fast but are not safe if their rider falls off the slide). To ensure that the rider is released from the same spot for each trial, have students put a piece of masking tape approximately 5 cm from the top of the ramp as their "starting line." Instruct students to simply release, not push, the rider.

1. Have student teams complete the first two steps of the Procedure section of the Ramp Investigation #1 handout ("1. Decide on your research question and list your materials" and "2. Use the first two columns of the data table to make your plan") and then stop to get your approval of their plan before moving on.

2. Review each team's plan to be sure that the slopes they have designated are realistic (i.e., they will be able to elevate the ramp using the materials they have designated: e.g., height designated for ramp is not larger than the length of the ramp).

3. After teams have received approval, they should gather their materials, set up the ramps, and begin testing.

4. Students will conduct tests to determine the best incline for their ramps using two different riders (toy cars and fabric-covered blocks). They will apply their findings to their slide design for the module challenge. (*They should conclude that steeper inclines allow the vehicle to move faster, but too steep an incline will cause the vehicle to drop off.*)

Ramp Investigation #2: Ramp Surface

Students experimented in Ramp Investigation #1 with a variety of slopes to guide their slide design. In Ramp Investigation #2, students will explore surfaces for their slides by experimenting with various surfaces on their ramps. Introduce the concept of friction as a force that works between objects rubbing against one another. When an object is traveling, friction is a force that works in the opposite direction of the object's direction of travel. Roll a ball across the floor as an example, pointing out to students that the ball does not continue to roll, but slows down and eventually stops. This is because of the friction between the ball and the floor. Ask students to share examples of friction. Next, ask students how friction is important for a slide (e.g., if there is too much friction, the rider won't be able to slide well; if there is too little friction the rider will go very fast and could be hurt).

As a class, discuss which rider from Ramp Investigation #1 (toy car or fabric-covered block) moves on the slide most like a person. Students should conclude that the fabric-covered block is more similar to a person than the toy car. Tell students that in this investigation they will need to use their findings about the fabric-covered block from Ramp Investigation #1. Each team should identify the slope on which their block traveled the farthest and use this ramp height for Ramp Investigation #2. Each team will experiment with wax paper, aluminum foil, sandpaper, and felt coverings for their ramps. In addition, each team should choose one other material from a supply you have on hand that they believe would make a good slide material (e.g., plastic wrap, wrapping paper, shelf liner, fabric).

Distribute the Ramp Investigation #2 handout to each student and investigation materials to each team. Students should complete the first step in the Procedure sections of the handout ("1. Decide on your research question and list your materials").

Students will conduct tests to determine the best surface for their slide. (*They should conclude that the materials that are the most slippery—e.g., wax paper, aluminum foil—offer the greatest amount of speed in the descent. The problem with metals, of course, is that they get very hot in the heat of the summer and very cold in the winter.*)

After teams have experimented with the various ramp coverings, have each team choose one covering to use in a class "race." Have all students release their riders and compare how far their riders traveled. Create a class table that lists the height of each ramp, the covering material, and the distance of travel. Hold a class discussion about the results, identifying the ramp height and covering that resulted in the fastest slide (as identified by the distance the rider traveled). Have students consider how they could apply their learning from this activity to a slide design in a STEM Research Notebook prompt.

STEM Research Notebook Prompt

Think about how the various ramps your class created performed. What ramp covering worked the best? What ramp height worked the best? How can you use this information to create a slide that has a high fun factor rating and is safe?

Slide Makeover Plan Book

Begin the activity by asking the class the following questions:

- If you owned a company that designed and built playgrounds for schools and parks, what kind of workers would you need to design and build awesome playgrounds? (*Introduce the idea that engineers are people who design and build things to solve problems or fulfill human needs. Students should understand that there are many different kinds of engineers and that these engineers do different kinds of work.*)

- How do engineers get new ideas for swing set designs? (*Engineers do a lot of research and experiments before they develop new designs, but even then they often build their model only to have to start again. Encourage students that when designing new ideas, they may begin with a slide idea they see online, but then tweak it to make it better.*)

As mentioned in the first bullet above, discuss with students that engineers are people who design and build things to solve problems or fulfill human needs. Students should understand that there are many different kinds of engineers (see the Lesson Plan 1 Teacher Background Information section on p. 53 for an overview) and that these engineers do different kinds of work.

Remind students that they will act as engineers in this module as they design and build a swing set. To assume the role of an engineer, they will need to solve problems and use the engineering design process, or EDP, to do their work.

Show the "Engineering Process" video from Crash Course Kids (*www.youtube.com/watch?v=fxJWin195kU*) to students to introduce the steps of the EDP. After the video, show students the EDP graphic attached at the end of the lesson and tell them that they will use this process as they work in groups to solve their Swing Set Makeover Design Challenge. Tell students that they will have the chance to practice using the steps of the EDP now and to put into action what they've learned about ramps.

Hold a class discussion about collaboration or teamwork, and emphasize to students that engineers, scientists, and other professionals often work in groups. Have students

share their ideas about working in teams, asking them to share their ideas about whether they like working in groups (why or why not), what makes teams work well together, and what their personal responsibilities are in groups. Start a class list of good collaboration practices. Provide each student with a copy of the Collaboration Rubric (p. 131). Students will evaluate themselves and other members of their team.

In this activity, students revisit the Fun Factor Survey from Lesson 1, are challenged to use what they learned from the ramp investigations, and integrate their fun factors with their findings about slide slope and coverings to create a new slide design that is both fun and safe.

The goal is for students to draw a sketch for a slide that they will build later in the module. Here are some questions the design teams may need to consider:

- If the ramp you designed in Ramp Investigation #2 were a part of the swing set, what would its fun factor rating be?

- What was most important to the fun factor rating: slope or surface?

- What changes could cause the rider to move faster (or slower if that is your team's wish)?

- What features could be added to make the slide better?

Instruct students to use the EDP to create a plan for a new slide using the Slide Makeover Plan Book handout (p. 129). Student teams should collaborate to work through the steps of the EDP; however, each student should create his or her own plan book using four sheets of paper and the directions on the handout. The Plan Book will include an annotated sketch. Tell students that they will create sketches for each component of their swing set makeovers. Review the Slide Design Sketch Rubric (p. 140) with students, emphasizing the importance of including detailed labels and explanations on their sketches. After students complete their plan books and sketches, teams should choose one sketch to use for their slide design. This may be one team member's sketch, or teams may choose to create a new sketch that incorporates elements of sketches from more than one team member.

After teams have selected one sketch for their slide design, have teams review one another's sketches and provide feedback, using the following format:

- I like the following about this slide design

- I think this slide design could be improved by

- I have questions about

Have teams reconvene and discuss the feedback they received. Teams should revise their slide design sketches based on that feedback.

Mathematics Connection: Introduce the idea that the shapes we see used in a building serve a particular purpose in the building's structure and help make the building strong and functional. Ask students whether they see more straight lines or curved lines in the structure of their classroom. Tell students that closed shapes made of straight lines that are connected are called polygons. Ask students to name shapes that are polygons (e.g., squares, triangles, hexagons) and shapes that are not polygons (e.g., circles, ovals, line segments), justifying their classification of shapes. Ask students to share their ideas about what the "strongest" shape is (i.e., the shape that could hold up the most weight without breaking). Students may be surprised to learn that the triangle is the strongest of the shapes. Point out to them the use of triangles to strengthen the barn door in the barn door image. Ask students to brainstorm ideas about why they think the triangle is so strong. Tell students that the forces in a triangle are evenly distributed among all three sides, making it very strong. Ask students for their ideas about things that are constructed of triangles (e.g., Egyptian pyramids). Tell students that they will explore their school's playground to look for shapes and many of these shapes are used to make the playground equipment strong, safe, and fun. Have students brainstorm a list of shapes they might see on the playground, creating a class list. You may wish to show a video that highlights shapes in structure such as the Design Squad video "Strong Structures With Triangles" (*www.youtube.com/watch?v=mBHJtWbsiaA*).

Next, ask students to imagine that they are able to hover over the playground in a helicopter, looking down at the playground from above. Ask students for their ideas about what they would see, and draw a picture on a whiteboard or on chart paper reflecting students' ideas. Guide students to understand that they would notice that the playground is situated within a defined area. The edges of this area may be defined by some sort of wood curbing or by a change in the type of surface (e.g., mulch on the playground and grass outside the playground). Draw edges onto your drawing, approximating the shape of your school playground.

Introduce the terms *perimeter* and *area* using the drawing you created. Tell students that the line representing the edges of the playground is the perimeter and the amount of space inside the edges of the playground is the area. Ask students for their ideas about how they would measure the perimeter of the playground and what units they would use to do this, guiding students to understand that the perimeter is the sum of the lengths of each edge. Next, ask students for their ideas about how they would measure the area of the playground and what units they would use to do this. Guide students to understand that the area of a rectangle is its length multiplied by its width and that its units will be in m^2 or ft^2, and ask students to share their ideas about why these "squared" units are used for area. *Note:* If your school's playground does not have a rectangular footprint, guide students to understand that by dividing the footprint into sections (e.g., a rectangle and a square), they can find the area of an irregular shape.

If students are unfamiliar with the concepts of perimeter and area, you may wish to give students additional practice with these concepts. For example, you may wish to have students or design teams create an anchor chart (see the Teaching Strategies section on p. 105 and the Explanation section on p. 118) for perimeter and area or to measure and calculate the perimeters and areas of various shapes they find in the classroom (e.g., desk or tabletops, windows, or the classroom footprint).

Geometry Scavenger Hunt Activity

Students should use the Geometry Scavenger Hunt handout (p. 133) to work with their design team to investigate the shapes present in their school playground and to apply their understanding about area and perimeter.

After teams have completed their Geometry Scavenger Hunts, have each student reflect on their findings in a STEM Research Notebook entry.

STEM Research Notebook Prompt

Ask students to consider their findings and report anything they found interesting or surprising. You may guide their reflections by posing questions such as the following:

- *Do the shapes and sizes affect the motion of the swing or slide? How?*

- *How could friction (surface friction and air resistance) be minimized by changing the design of the swings and slides?*

- *What materials is the playground equipment made of?*

- *How are materials important to the fun factor of a playground?*

- *How are materials important to the safety of playground equipment?*

ELA Connection: Hold a class discussion about talking and writing as forms of communication, using questions such as the following:

- How is communicating by writing different from communicating by talking?

- Is it possible to change what someone thinks about a subject by writing about it? Why or why not?

- If you were going to write a blog or newsletter to convince someone to think about an issue in the same way you do, what would be some important things to consider?

Explain to students that they will create a blog or newsletter about building their swing set models. In the blog, they will try to convince their audience (a school that wants to purchase a new swing set) that the swing set they designed is both fun and safe.

Teams will practice writing persuasive arguments after looking at swing sets online and reading reviews. This activity will help teams defend their design decisions in their blogs and in their presentations of their swing set designs.

Ask the students if they have ever eaten an Oreo cookie. Tell them that the letters that make up the word *OREO* can help them remember the steps in writing an opinion paper. Pass out the Writing the OREO Way handout and review the components of opinion writing outlined on the handout. The OREO writing steps are as follows:

- O = Write an introduction—clearly state your *opinion.*

- R = Give two or three *reasons* why you hold this opinion.

- E = Provide two or three detailed *examples* to support these reasons.

- O = Restate your *opinion* in your conclusion.

Remind students that there are no right or wrong opinions, but that opinions should be supported with reasons (evidence).

Pass out (or have teams review online) three or four swing set design reviews. Based on the reviews (these will be used as evidence for teams' opinion writing), ask each team to choose a design they would purchase. Have teams apply the OREO strategy to write short opinion papers about why they would purchase one of the reviewed swing sets. Teams should also create a graphic organizer, a brochure, or an oral presentation highlighting reasons why they would purchase the swing set.

Park Exhibition Showcase Activity

Social Studies Connection: Have students finish up the Park Exhibition Showcase research and presentation. Review expectations by reviewing the Park Exhibition Showcase handout (p. 87) and Park Presentation Rubric (p. 91) provided in Lesson 1.

Allow students to share their findings with the class and answer questions. During the presentation students should do the following:

- Highlight the most important features of the park.

- Show the artifacts they found.

- Describe the play spaces at the park and characteristics about the playground that they particularly liked or disliked.

- Provide an opportunity for peers to ask questions and share experiences at that park.

Make arrangements with the school librarian to set up the displays in the library with all or the best of the exhibits as a showcase.

Explanation

Science Class: For the Ramp Investigation activities, check to make sure the students release the vehicles from the top of the ramp in the same spot every time. Also, explain that the ramp should be lifted with uniform intervals to enable the teams to make reasonable predictions about the results (e.g., lift using one textbook at a time). Emphasize the importance of collaboration in projects (you may wish to refer to engineers and how they work).

The following literature connections may be useful to reinforce science concepts in this lesson:

- *Roll, Slope, and Slide: A Book About Ramps* by Michael Dahl

- *Roller Coaster* by Marla Frazee

Mathematics Connection: Understanding area and perimeter and keeping the terms straight can be difficult for students. Students can create a graphic organizer called an anchor chart for the two terms and keep it inside their STEM Research Notebook. To create the anchor charts, have students divide a piece of paper into three columns and label the columns "is," "can," and "looks like." One example follows:

- Column 1: Perimeter *is*/Area *is*. Define the terms using a mathematics book definition, then define it again using student's own words.

- Column 2: Perimeter *can*/Area *can*. How can these concepts be used in everyday life outside of the mathematics classroom? Students provide two or more examples.

- Column 3: Perimeter looks *like*/Area looks *like*. Students create diagrams and mathematics problems and show how to solve the mathematics equation visually.

Students are asked to identify acute, obtuse, and right angles as a part of the Geometry Scavenger Hunt. You may wish to review these concepts with students before beginning the scavenger hunt.

ELA Connection: You may wish to provide an example of an opinion essay for students to read before teams create their opinion papers. An example of such an essay is *I Want a Dog: My Opinion Essay* by Darcy Pattison.

Remind students to use good writing practices as teams create their OREO opinion papers. Have students brainstorm to create a class list of good writing practices such as the following:

- Use linking words and phrases to connect ideas.

- Organize writing into paragraphs.

- Use capitalization and punctuation.

- Spell words correctly.

Social Studies Connection: Tell students that the rights we have as citizens come with the responsibility to help others in our community. Review the meaning of responsibility and associate the term with chores they do or other members of the family do at home. Ask students to consider what happens when chores are not done. Discuss with the students that our local community is like a very large family. Discuss what it means to be a good citizen.

Elaboration/Application of Knowledge

Science Class: Have students make a list of ramps they see in their neighborhood and on their way to school and identify how the ramps are used.

Hold a class discussion about technology, launching the discussion by asking students to share their ideas about what technology is. Formulate a class definition of the term and have students offer examples of technologies they see every day (e.g., computers, cars). Ask students if playgrounds incorporate technology. Introduce the idea that various features of playground equipment are technologies, emphasizing that a technology is a way that humans apply science to use objects from their environments in new ways to serve a human purpose. Explain to students that technologies such as swings and ladders are so common in our society that we do not think of them as technologies. Distribute one swing set image from Lesson 1 to each team and have students work in their teams to identify technologies in each image. If students have difficulty understanding how technology is incorporated into swing sets, you may wish to compare the modern swing set images with images of older playground equipment to identify features of modern swing sets that are improvements on the older equipment.

Once students can identify technological innovations, this understanding can be extended to introduce the idea that technology is apparent in many aspects of playground equipment, including the fasteners used to hold the equipment together, the hinges used on swinging parts of the equipment, the metals used in their construction, and the application of scientific knowledge about ramps and pendulums to create slides and swings.

Mathematics Connection: Students can practice measuring and creating diagrams using their measurements by measuring the furniture in a room of their home and creating an annotated map of the room on graph paper.

ELA Connection: Have students share their opinion papers with a family member and report back the family member's opinion on the topic and evidence provided.

Have students choose some of their favorite books and read online reviews about them (positive and negative). Ask the students to share their ideas via a think/pair/share activity: First, students should reflect individually on their ideas about how the review affects how people perceive this product. Next, students should pair with a partner to share these ideas. Ask them to share whether the review would influence their own purchasing decision.

Social Studies Connection: Not applicable.

Evaluation/Assessment

Students may be assessed on the following performance tasks and other measures listed.

Performance Tasks

- Ramp investigations
- Slide Makeover Plan Book and sketch
- Geometry Scavenger Hunt
- Opinion essay
- Park Presentation activity

Other Measures

- Engagement in class activities and discussions
- Involvement in group work and discussions
- STEM Research Notebook entries

INTERNET RESOURCES

"Balanced and Unbalanced Forces" video from the Fuse School
- *www.youtube.com/watch?v=YyJSlcIbd-s*

Information on different engineering jobs
- *www.sciencekids.co.nz/sciencefacts/engineering/typesofengineeringjobs.html*

Your Weight on Other Worlds web page
- *www.exploratorium.edu/ronh/weight*

Genia Connell's blog on opinion writing
- *www.scholastic.com/teachers/blog-posts/genia-connell/graphic-organizers-opinion-writing*

"750 Feet Sand Marble Race" video

- *www.youtube.com/watch?v=UaMr6bFruF0*

"The Engineering Process: Crash Course Kids" video

- *www.youtube.com/watch?v=fxJWin195kU*

"Strong Structures With Triangles" video from the Design Squad

- *www.youtube.com/watch?v=mBHJtWbsiaA*

Videos to inspire student slide designs

Water slides

- *www.youtube.com/watch?v=hhk2uqz5umA*

- *www.youtube.com/watch?v=JA8iBapiQRo*

Straight slides

- *www.youtube.com/watch?v=r56AHblQvlo*

- *www.youtube.com/watch?v=jmtWo9nRLEE*

- *www.youtube.com/watch?v=iio5zXVgiUI*

- *www.youtube.com/watch?v=eRh4DVbMcNY*

Curved and spiral slides

- *www.youtube.com/watch?v=U-q_Qvn0iTc*

- *www.youtube.com/watch?v=KeLpPljQCK4*

- *www.youtube.com/watch?v=15oG0yU8kcU*

Tunnel slides

- *www.youtube.com/watch?v=KBP7LMfhB6k*

- *www.youtube.com/watch?v=jM3WqbY3glY*

Name: _____   Team Name: _____

RAMP INVESTIGATION #1: RAMP HEIGHT

OVERVIEW

You will do an experiment that will help you understand the motion and forces of an object moving down a ramp. The questions you will answer should help you design an extremely fun slide that is also safe.

Your first rider is a toy car. After you complete your trials with the car, repeat the procedure with a fabric-covered block. Your rider must stay on the slide during the entire ride; if your rider falls off, the distance for that trial will be 0.

You will measure how fast your rider moves down the ramp by measuring how far the rider goes after it leaves the end of the slide (measure from the end of the ramp to the front end of the car or block after it stops moving).

To create different ramp slopes, use materials from the classroom to lift one end of the ramp to different heights. To be sure that the rider is released from the same spot for each trial, put a piece of masking tape about 5 cm from the top of the ramp as the "starting line." Gently release (do not push) your rider.

MATERIALS

- Ramp
- Toy car and fabric-covered block
- Masking tape
- Meter stick
- Classroom materials for lifting one end of the ramp to different heights

SAFETY NOTES

1. All laboratory occupants must wear safety glasses or goggles during all phases of this inquiry activity.

2. Make sure there are no fragile materials in the area where activities are taking place.

3. Have an appropriate level of adult supervision to ensure safe behaviors during activities.

4. Use caution when working around slip/fall hazards (e.g., marbles, ramps).

5. Make sure all materials are put away after completing the activity.

6. Wash hands with soap and water after completing this activity.

Name: _____   Team Name: _____

RAMP INVESTIGATION #1: RAMP HEIGHT

PROCEDURE

1. Decide on your research question and list your materials:

   - Our research question for Ramp Investigation #1 is

   - We will use the following materials:

2. Use the first two columns of the data table to make your plan (enter the different heights of the ramp and the materials you will use to lift the ramp).

3. Put a piece of masking tape about 5 cm from the top of the ramp as the "starting line."

4. Set up your ramp for the first height.

5. Run three trials using the car. Record the distances for your car by measuring from the end of the ramp to the location of the front end of the car after it stops moving.

6. Run three trials using the fabric-covered block. Record the distances for your block.

7. Set up your ramp for the second height and repeat Steps 4 and 5.

8. Repeat until you have tested all the ramp heights you planned.

Name: _____   Team Name: _____

STUDENT HANDOUT, PAGE 3

RAMP INVESTIGATION #1: RAMP HEIGHT

DATA TABLE

Ramp Height	Materials Used to Lift Ramp	Distance Car Travels			Distance Block Travels		
		Trial 1	Trial 2	Trial 3	Trial 1	Trial 2	Trial 3

INVESTIGATION QUESTIONS

1. At what ramp height did your car travel the farthest?

2. At what ramp height did your block travel the farthest?

3. Which traveled the farthest—the car or the block? Why do you think this happened?

4. Were any of the ramp heights unsafe (i.e., your rider fell off)?

5. How will you use the information from this investigation to design your swing set's slide?

6. What other ideas do you have that you might want to test?

④

Name: _____   Team Name: _____

RAMP INVESTIGATION #2: RAMP SURFACE

OVERVIEW

Your team is working to build the best slide! In this investigation, you will experiment with coverings for your ramp. You will experiment with wax paper, aluminum foil, sandpaper, and felt to see what covering makes your rider travel the farthest. You will also choose one other material from the classroom supply to try.

MATERIALS

- Masking tape
- Meter stick
- Scissors
- Fabric-covered block
- Wax paper
- Aluminum foil
- Sandpaper
- Felt
- One other material from the classroom supply

SAFETY NOTES

1. All laboratory occupants must wear safety glasses or goggles during all phases of this inquiry activity.

2. Make sure there are no fragile materials in the area where activities are taking place.

3. Have an appropriate level of adult supervision to ensure safe behaviors during activities.

4. Use caution when working around slip/fall hazards (e.g., marbles, ramps).

5. Make sure all materials are put away after completing the activity.

6. Wash hands with soap and water after completing this activity.

Name: _____   Team Name: _____

RAMP INVESTIGATION #2: RAMP SURFACE

PROCEDURE

1. Decide on your research question and list your materials:

   • Our research question for Ramp Investigation #2 is

   • We will use the following materials:

2. Build your ramp using the best incline (the one for which the fabric-covered block traveled the farthest in Ramp Investigation #1).

3. Choose a material to cover your ramp. Cut it to a size about 2 inches wider than your ramp (or tape together pieces to make it long enough to cover your ramp).

4. Cover the ramp with the material, folding it over the edges of your ramp and taping it to the back with masking tape.

5. Mark a spot 5 cm from the top as your "starting line."

6. Conduct three trials, measuring how far your rider travels each time.

7. Record the distances in the data table.

8. Repeat Steps 3–7 for the wax paper, aluminum foil, sandpaper, and felt.

9. Choose one other material from the classroom supply that you think will create a fast sliding surface.

10. Repeat Steps 3–7 for this material.

Name: _____   Team Name: _____

RAMP INVESTIGATION #2: RAMP SURFACE

DATA TABLE

Material	Trial 1 Distance	Trial 2 Distance	Trial 3 Distance

INVESTIGATION QUESTIONS

1. What surface created the fastest ride?

2. What surface created the slowest ride?

3. How can you use this information to create your slide?

4. Think about safety. Might any of these materials be unsafe? For example, think about what happens to metal when it is left in the sun.

5. What other ideas could you test?

ENGINEERING DESIGN PROCESS

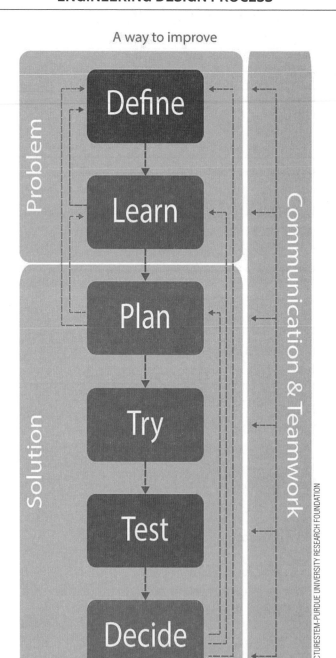

Name: _____   Team Name: _____

Our swing set name is _____

SLIDE MAKEOVER PLAN BOOK

Now it is time to begin thinking like an engineer! Use the following tables to help you work through your team's slide design like an engineer. You will need four pieces of paper. Put the pages together and fold them in half, like a book. Give it a title, then turn the page and begin. You will have two pages of your book for each of the first three tasks. You will create your drawing (the fourth task) on graph paper and then glue or staple it into your book.

TASK 1: DEFINE THE PROBLEM—FIND THE IMPORTANT PARTS

Description What is the role of a slide on the swing set?	**Fun Factor Strong Points** What fun factor elements are met by a slide's role?
Safety What safety rules apply to slides?	**Other Things to Consider** What other limits might be important?

TASK 2: IMAGINE IT!—BRAINSTORM POSSIBLE SOLUTIONS

Ideas to Add Function How can you change the slide to add new ways to play on it?	**Ideas for Fun** What can you do to the slide to make it more fun?
Safety Pointers How could you make the slide even safer?	**More Ideas to Improve the Slide** What other ideas do you have?

Name: _____   Team Name: _____

Our swing set name is _____

SLIDE MAKEOVER PLAN BOOK

TASK 3: PLAN IT!—RESEARCH, LIST MATERIALS, AND IDENTIFY THE NEXT STEPS

Functions: Research It!	Fun: Research It!
What have others done to add functions to the basic slide design?	How have others made the slide more fun?
Safety: Research It!	**More Ideas: Research It!**
What are some of the rules for safe slide designs?	What else did you learn when you were conducting your research?

TASK 4: SKETCH IT!—FOLLOW YOUR PLAN AND DRAW YOUR SLIDE DESIGN

Make a sketch of your slide design on graph paper. Label all of your slide's parts, including the height and width, and list the materials you will use to build it. Add notes about anything that will help make your sketch easy to understand when you use it to build your slide. After you complete your sketch, glue or staple it into your plan book.

TASK 5: SHARE IT!—ENGINEERS SHARE THEIR IDEAS AND USE FEEDBACK

Share your sketch with your team members. Compare all of the team members' sketches and discuss what you like about each one. Choose one sketch to use as your team's slide design, or create a new sketch using things you like from more than one sketch.

Collaboration Rubric

Name: _____

Expectations	Needs Improvement (1 point)	Approaching Standard (2 points)	Meets Standard (3 points)	Exceeds Standard (4 points)	Score
SELF-REFLECTION I stayed on task while working with my group.					
SELF-REFLECTION I communicated and recorded our thoughts.					
SELF-REFLECTION I followed directions for the activity.					
SELF-REFLECTION I completed the task successfully with my group.					
SELF-REFLECTION I did an equal amount of work, the same as the others in my group.					
PEER EVALUATION Participated in group research.					
PEER EVALUATION Did an equal share of the work.					
PEER EVALUATION Had a positive attitude about the project.					
TEACHER INPUT Student has an equal share of research and contributed an equal number of ideas.					
TEACHER INPUT Student was able to report valuable findings to the group.					

TOTAL SCORE: _____

COMMENTS:

BARN DOOR IMAGE

Name: _____   Team Name: _____

GEOMETRY SCAVENGER HUNT

SAFETY NOTES

1. Use caution when working out in the field as there can be several trip/fall or slip/fall hazards (e.g., sports equipment, uneven ground, holes) that can cause physical injuries.

2. Make sure there are no fragile materials in the area where activities are taking place.

3. Have an appropriate level of adult supervision to ensure safe behaviors during activities.

4. Use caution when working with sharps (e.g., scissors, sticks) to avoid cutting or puncturing skin or eyes.

5. Make sure all materials are put away after completing the activity.

6. Wash hands with soap and water after completing this activity.

DIRECTIONS

Assign the following roles to team members:

- Measurer (hold the tape measure and call out the measurements to the recorder)

- Recorder (hold the clipboard and record the information the team collects)

- Mathematician (hold the calculator and make any calculations needed for the team)

- Quality Control (read question to the team, help the measurer, and double check answers)

Your team will need a clipboard, a pencil, a tape measure, and a calculator for this activity.

Be sure to measure carefully as you answer the questions. The answers will provide important information for you when you create your swing set later.

Name: _____   Team Name: _____

STUDENT HANDOUT, PAGE 2

GEOMETRY SCAVENGER HUNT

DISTANCE AND AREA OF THE PLAYGROUND

1. Measure all the edges of the playground to find the distance around it. Record the distances you measured for each edge here:

   _____   _____   _____   _____

   - The total distance around the edges of the playground is the sum of the distance of each edge. This is called the **perimeter.** The perimeter of the playground is _____

2. Compare your perimeter result with another team. Did you get the same answer? _____

   - If you did not get the same answer, make your measurements again to check them.

3. Multiply the length of the playground by the width ($l \times w$) to find the **area** of the playground: _____

Name: _____   Team Name: _____

STUDENT HANDOUT, PAGE 3

GEOMETRY SCAVENGER HUNT

EXPLORING SHAPES

4. Identify items on the swing set that have the same or similar shapes:

5. Identify two quadrilaterals (shapes with four sides). What shapes are these and where did you find them?

6. Identify two triangles and where you found them:

7. Is there a cylinder on the playground? Where did you find it?

8. Locate and name two angles (acute angle, obtuse angle, or right angle).

- Angle #1: _____

- Angle #2: _____

Name: _____   Team Name: _____

GEOMETRY SCAVENGER HUNT

SWING MEASUREMENTS

9. What is the area of one of the quadrilaterals you found ($l \times w$) _____

10. How long is the chain on the swing? _____

11. How far off the ground is the swing? _____

SLIDE MEASUREMENTS

12. What is the height of the slide? _____

13. Take some more measurements: How long is the sliding surface of the slide? _____

14. How wide is the slide? _____

15. How long is the straight part at the end of the slide? _____

16. How far off the ground is the end of the slide? _____

Name: _____   Team Name: _____

GEOMETRY SCAVENGER HUNT

MOTION MYSTERIES

17. Where on the playground equipment do you move vertically (up and down)?

18. Where on the playground equipment do you move horizontally (sideways)?

19. Where on the playground equipment do you move in a straight line?

20. Where on the playground equipment do you move in a circle?

Name: _____

STUDENT HANDOUT

WRITING THE OREO WAY

Opinion	Give your opinion.
Reason	State the reason for having this opinion.
Example	Give an example that supports your opinion.
Reason	State a second reason for having this opinion.
Example	Give a second example that supports your opinion.
Opinion	Restate your opinion.

Name: _____

Ramp Investigations Rubric

Criteria	Needs Improvement (1 point)	Approaching Standard (2 points)	Meets Standard (3 points)	Exceeds Standard (4 points)	Score
RESEARCH QUESTION	The question is not testable and is not related to the investigation.	The question is testable, but is not related to the investigation.	The question is testable and is related to the investigation.	The question is testable and is directly related to the investigation.	
MATERIALS	More than four materials are missing.	Three or four materials are missing.	Most materials are listed but one or two may be missing.	All materials are listed.	
EXPERIMENTAL DESIGN	More than four procedures are missing and some key terms are not used correctly.	Three or four procedures are missing, but key terms are used correctly.	Most procedures are listed and most key terms are used correctly.	All procedures are listed in the correct order and key terms are used correctly.	
DATA AND RESULTS	Much of the data are not recorded and it is not possible to find a pattern.	Data are shown in such a way that it is not possible to make clear predictions.	Some data are shown so that some patterns can be seen.	Data are shown so that any patterns can be seen.	
CONCLUSION	Conclusion is missing or includes only one of the following: • Results of the tests • How this relates to the Swing Set Makeover Design Challenge • What was learned	Conclusion includes at least two of the following: • Results of the tests • How this relates to the Swing Set Makeover Design Challenge • What was learned	Conclusion includes all of the following: • Results of the tests • How this relates to the Swing Set Makeover Design Challenge • What was learned	Conclusion includes all of the following: • Results of the tests • How this relates to the Swing Set Makeover Design Challenge • What was learned • Ideas for future testing	
ACCURACY	Descriptions of scientific terms and facts are incorrect or missing.	Descriptions of scientific terms and facts are partly complete and accurate.	Descriptions of scientific terms and facts are mostly complete and accurate.	Descriptions of scientific terms and facts are complete and accurate.	

TOTAL SCORE: _____

COMMENTS:

Slide Design Sketch Rubric		
Name: _____	Team Name: _____	
Points	**Standards**	**Details**
25	• Detailed notes describe the parts of the slide (e.g., measurements, materials). • The slide design is practical (i.e., if the design were used, a working slide would result). • The slide design clearly reflects understanding of ramps and gravity. • Reasoning for material choices is provided and clearly explained. • The design is unique and creative.	• The sketch is neat and may be colored with colored pencils. • A complete materials list is provided.
20	• Notes help describe the parts of the slide. • The slide design is practical (i.e., if the design were used, a working slide would result). • The slide design reflects understanding of ramps and gravity. • Some reasoning for material choices is provided. • The design is different from the examples shown.	• The sketch is somewhat neat and may be colored with crayons. • A materials list is provided.
10	• Some notes are provided but are not very helpful. • It is not clear that the slide design is practical (i.e., if the design were used, a working slide may or may not result). • The slide design reflects limited understanding of ramps and gravity. • Reasoning for material choices is provided but is unclear. • The design is similar to the examples shown.	• The sketch is not as neat as it could be and does not include color. • The materials list is incomplete.
5	• Sketch does not have notes to help explain the features of the slide. • The slide design is impractical (i.e., if the design were used, a working slide would not result). • The slide design reflects little or no understanding of ramps and gravity. • No reasoning for material choices is provided. • The design is a copy of the examples shown.	• The sketch is not neat and is not colored. • No materials list is provided.

SCORE: _____

COMMENTS:

Lesson Plan 3: Swinging Pendulums

In this lesson, students continue to explore the forces acting on swing set equipment and the motion that results. Science activities focus on balanced and unbalanced forces as demonstrated by pendulums. Lessons learned from measuring pendulum motion are applied to observations of swings. These observations will, in turn, be used to design the swing component of the swing set. Mathematics activities include measuring pendulum motion as well as ranking design alternatives using the rubric the teams created in the Fun Factor Survey handout from Lesson 1. ELA activities focus on preparing and evaluating arguments for proposed designs. In social studies, students discuss the responsibilities of a citizen and evaluate economic arguments based on resource constraints.

ESSENTIAL QUESTIONS

- What forces are at work on a playground swing?
- How does the design of the swing unit affect speed and motion?
- How do gravity and inertia affect the fun factor of a swing?
- How do materials used in a swing unit affect the cost or performance of the unit?
- What safety features are part of the design of a swing?

ESTABLISHED GOALS AND OBJECTIVES

At the conclusion of this lesson, students will be able to do the following:

- Recognize and describe gravity and inertia's influence on pendulum motion
- Analyze the motion of a pendulum to make predictions
- Predict and evaluate the impact of design and materials on a swing's fun factor rating
- Understand that components of swing design include motion, safety, materials, and aesthetics
- Compare and contrast materials and pendulum lengths to make decisions for a new swing design.
- Design a swing for the Swing Set Makeover Design Challenge

TIME REQUIRED

- 5 days (approximately 45 minutes each day; see Table 3.8, p. 42)

MATERIALS

Handouts for Lesson 3

- Pendulum Investigation #1 and #2 (1 of each per team)
- Jigsaw Research Template (1 per team)
- Swing Makeover Plan Book (1 per team)

Rubrics for Lesson 3

- Pendulum Investigations Rubric
- Swing Design Sketch Rubric

Necessary Materials for Lesson 3

- STEM Research Notebooks (1 per student)
- The book *Follow Me* by Tricia Tusa (HMH Books for Young Readers, 2011)
- The poem "The Swing" by Robert Louis Stevenson
- Internet access (for showing video clips and for student research)
- KWL chart from previous weeks
- Optional: pictures of swing sets with swings and other forms of pendulums

Additional Materials for Pendulum Investigations

- 4 full-length pencils
- 4 pieces of string—approximately 30 cm, 45 cm, 60 cm, and 75 cm
- 1 roll of masking tape—approximately 1-inch wide
- 10 large washers, identical in size and mass—approximately 3 cm in outside diameter (a pipe cleaner will be wrapped around the washers to hold them together in groups)
- 4 pipe cleaners (or chenille stems)
- 4 meter sticks
- 4 stopwatches
- 4 calculators
- 4 colored markers or pencils—one each of four different colors. Each station is identified by its color.

- 1 scale that can measure grams
- Safety glasses or goggles

Additional Materials for Jigsaw Research Activity

- Internet access
- Chart paper (per team)
- Markers (4 different colored markers per team)

SAFETY NOTES

1. All laboratory occupants must wear safety glasses or goggles during all phases of this inquiry activity.

2. Make sure there are no fragile materials in the area where activities are taking place.

3. Have an appropriate level of adult supervision to ensure safe behaviors during activities.

4. Use caution when working with pendulums:

   - It is helpful to have a clear idea for setting up the pendulums so students have plenty of room for swinging the pendulum without hitting team members. The longest string is approximately 30 inches.

   - Students should be cautioned to carefully drop the pendulum bob without applying additional force.

   - At no time should the pendulum bob be held in a student's hand and swung through the air.

   - Students should always wear safety glasses or goggles when working with or observing flying objects.

5. Make sure all materials are put away after completing the activity.

6. Wash hands with soap and water after completing this activity.

CONTENT STANDARDS AND KEY VOCABULARY

Table 4.10 lists the content standards from the *NGSS, CCSS,* and the Framework for 21st Century Learning that this lesson addresses, and Table 4.11 (p. 148) presents the key vocabulary. Vocabulary terms are provided for both teacher and student use. Teachers may choose to introduce some or all of the terms to students.

Table 4.10. Content Standards Addressed in STEM Road Map Module Lesson 3

NEXT GENERATION SCIENCE STANDARDS

PERFORMANCE EXPECTATIONS

- 3-PS2-1. Plan and conduct an investigation to provide evidence of the effects of balanced and unbalanced forces on the motion of an object.

- 3-PS2-2. Make observations and/or measurements of an object's motion to provide evidence that a pattern can be used to predict future motion.

- 3-5-ETS1-1. Define a simple design problem reflecting a need or a want that includes specified criteria for success and constraints on materials, time, or cost.

- 3-5-ETS1-2. Generate and compare multiple possible solutions to a problem based on how well each is likely to meet the criteria and constraints of the problem.

SCIENCE AND ENGINEERING PRACTICES

Asking Questions and Defining Problems

- Ask questions about what would happen if a variable is changed.

- Ask questions that can be investigated and predict reasonable outcomes based on patterns such as cause and effect relationships.

- Define a simple design problem that can be solved through the development of an object, tool, process, or system and includes several criteria for success and constraints on materials, time, or cost.

Developing and Using Models

- Collaboratively develop and/or revise a model based on evidence that shows the relationships among variables for frequent and regular occurring events.

- Develop a model using an analogy, example, or abstract representation to describe a scientific principle or design solution.

- Develop a diagram or simple physical prototype to convey a proposed object, tool, or process.

- Use a model to test cause and effect relationships or interactions concerning the functioning of a natural or designed system.

Planning and Carrying Out Investigations

- Plan and conduct an investigation collaboratively to produce data to serve as the basis for evidence, using fair tests in which variables are controlled and the number of trials considered.

Continued

Table 4.10. (*continued*)

- Make observations and/or measurements to produce data to serve as the basis for evidence for an explanation of a phenomenon or test a design solution.

- Make predictions about what would happen if a variable changes.

Analyzing and Interpreting Data
- Represent data in tables and/or various graphical displays (bar graphs, pictographs, and/or pie charts) to reveal patterns that indicate relationships.

- Analyze and interpret data to make sense of phenomena, using logical reasoning, mathematics, and/or computation.

- Compare and contrast data collected by different groups in order to discuss similarities and differences in their findings.

Using Mathematics and Computational Thinking
- Organize simple data sets to reveal patterns that suggest relationships.

- Describe, measure, estimate, and/or graph quantities (e.g., area, volume, weight, time) to address scientific and engineering questions and problems.

- Create and/or use graphs and/or charts generated from simple algorithms to compare alternative solutions to an engineering problem.

Constructing Explanations and Designing Solutions
- Use evidence (e.g., measurements, observations, patterns) to construct or support an explanation or design a solution to a problem.

- Identify the evidence that supports particular points in an explanation.

- Apply scientific ideas to solve design problems.

- Generate and compare multiple solutions to a problem based on how well they meet the criteria and constraints of the design solution.

Engaging in Argument From Evidence
- Compare and refine arguments based on an evaluation of the evidence presented.

- Distinguish among facts, reasoned judgment based on research findings, and speculation in an explanation.

- Respectfully provide and receive critiques from peers about a proposed procedure, explanation, or model by citing relevant evidence and posing specific questions.

- Construct and/or support an argument with evidence, data, and/or a model.

- Use data to evaluate claims about cause and effect.

- Make a claim about the merit of a solution to a problem by citing relevant evidence about how it meets the criteria and constraints of the problem.

Continued

Table 4.10. (*continued*)

Obtaining, Evaluating, and Communicating Information

- Read and comprehend grade-appropriate complex texts and/or other reliable media to summarize and obtain scientific and technical ideas and describe how they are supported by evidence.

- Compare and/or combine across complex texts and/or other reliable media to support the engagement in other scientific and/or engineering practices.

- Combine information in written text with that contained in corresponding tables, diagrams, and/or charts to support the engagement in other scientific and/or engineering practices.

- Obtain and combine information from books and/or other reliable media to explain phenomena or solutions to a design problem.

- Communicate scientific and/or technical information orally and/or in written formats, including various forms of media as well as tables, diagrams, and charts.

DISCIPLINARY CORE IDEAS

PS2.A: Forces and Motion

- Each force acts on one particular object and has both strength and a direction. An object at rest typically has multiple forces acting on it, but they add to give zero net force on the object. Forces that do not sum to zero can cause changes in the object's speed or direction of motion.

- The patterns of an object's motion in various situations can be observed and measured; when that past motion exhibits a regular pattern, future motion can be predicted from it.

PS2.B: Types of Interactions

- Objects in contact exert forces on each other.

CROSSCUTTING CONCEPTS

Cause and Effect

- Cause and effect relationships are routinely identified, tested, and used to explain change.

Patterns

- Patterns of change can be used to make predictions.

- Patterns can be used as evidence to support an explanation.

Structure and Function

- Different materials have different substructures, which can sometimes be observed.

- Substructures have shapes and parts that serve functions.

Influence of Science, Engineering, and Technology on Society and the Natural World

- People's needs and wants change over time, as do their demands for new and improved technologies.

- Engineers improve existing technologies or develop new ones to increase their benefits, to decrease known risks, and to meet societal demands.

Continued

Table 4.10. (*continued*)

COMMON CORE STATE STANDARDS FOR MATHEMATICS

MATHEMATICAL PRACTICES

- MP1. Make sense of problems and persevere in solving them.
- MP2. Reason abstractly and quantitatively.
- MP4. Model with mathematics.
- MP5. Use appropriate tools strategically.
- MP7. Look for and make use of structure.

MATHEMATICAL CONTENT

- 3.MD.C.5. Recognize area as an attribute of plane figures and understand concepts of area measurement.
- 3.MD.C.6. Measure areas by counting unit squares (square cm, square m, square inches, square feet, and improvised units).
- 3.MD.C.7. Relate area to the operations of multiplication and addition.
- 3.OA.B.5. Apply properties of operations as strategies to multiply and divide.
- 3.OA.D.9. Identify arithmetic problems (including patterns in the addition table or multiplication table), and explain them using properties of operations.

COMMON CORE STATE STANDARDS FOR ENGLISH LANGUAGE ARTS

READING STANDARDS

- RI.3.5. Use text features and search tools (e.g., key words, sidebars, hyperlinks) to locate information relevant to a given topic efficiently.
- RI.3.7. Use information gained from illustrations (e.g., maps, photographs) and the words in a text to demonstrate understanding of the text (e.g., where, when, why, and how key events occur).
- RI.3.10. By the end of the year, read and comprehend informational texts, including history/social studies, science, and technical texts, at the high end of the grades 2–3 text complexity band independently and proficiently.

WRITING STANDARDS

- W.3.1. Write opinion pieces on topics or texts, supporting a point of view with reasons.
- W.3.1.B. Provide reasons that support the opinion.
- W.3.2. Write informative/explanatory texts to examine a topic and covey ideas and information clearly.
- W.3.2.B. Develop the topic with facts, definitions, and details.
- W.3.3. Write narratives to develop real or imagined experiences or events using effective technique, descriptive details, and clear event sequences.

Continued

Table 4.10. (continued)

SPEAKING AND LISTENING STANDARDS

- SL.3.1. Engage effectively in a range of collaborative discussions (one-on-one, in groups, and teacher-led) with diverse partners on grade 3 topics and texts, building on others' ideas and expressing their own clearly.

- SL.3.1.D. Explain their ideas and understanding in light of the discussion.

- SL.3.4. Report on a topic or text, tell a story, or recount an experience with appropriate facts and relevant, descriptive details, speaking clearly at an understandable pace.

- SL.3.6. Speak in complete sentences when appropriate to task and situations in order to provide requested detail or clarification.

FRAMEWORK FOR 21ST CENTURY LEARNING

- Interdisciplinary Themes: Health and Safety; Environmental Literacy; Science; Mathematics

- Learning and Innovation Skills: Creativity and Innovation; Critical Thinking and Problem Solving; Communication and Collaboration

- Information, Media, and Technology Skills: Information Literacy; Media Literacy

- Life and Career Skills: Flexibility and Adaptability; Initiative and Self-Direction; Social and Cross-Cultural Skills; Productivity and Accountability; Leadership and Responsibility

Table 4.11. Key Vocabulary for Lesson 3

Key Vocabulary	Definition
arc	an unbroken part of a circle or other curved line
bob	the weight at the end of the pendulum
circle	a line that is curved so that its ends meet and every point on the line is the same distance from the center
footprint	the shape and size of the area an object occupies on the ground
frequency	the number of swings per minute
Galileo	an Italian astronomer and mathematician who was the first to use a telescope to study the stars
kinetic energy	the energy of motion
landscape architect	a person who designs the outdoor environment, especially parks or gardens
materials	the matter from which something is made
materials scientist	a person who studies the properties and use of things

Continued

Table 4.11. (*continued*)

Key Vocabulary	Definition
oscillation	one swing of an object that swings back and forth, like a pendulum
path	the way or track in which something moves
pendulum	a device made of weight hung from a string or rod so that the weight swings freely back and forth under the action of gravity
period	the time for one complete cycle; a back-and-forth motion of a pendulum
propulsion	the action of driving or pushing forward
pumping	when swinging, leg motion that increases potential energy; accomplished by extending the legs to the front during the forward-moving part of the swing's arc and tucking the legs back during the backward-moving part of the swing's arc
rhythm	a strong, regular, repeated pattern of movement
rotate	the action of turning around a center line or center point
scaled drawing	a drawing of an object that is either reduced or enlarged in size by a specific amount called a scale
stall	to bring to a standstill by stopping an object's motion
swing	a seat held up by ropes or chains on which someone may sit

TEACHER BACKGROUND INFORMATION

Students apply learning from previous lessons throughout the module in this lesson. You may wish to review the principles of force and motion covered in Lessons 1 and 2, particularly the concepts of gravity and inertia. This lesson includes a pendulum activity that will give students experience with a new type of motion that does not conform to the straight line inertia concept covered in Lesson 2. Students will also use the criteria they incorporated in the Fun Factor Survey from Lesson 1 as they design a swing unit for their Swing Set Makeover Design Challenge.

Pendulums

A *pendulum* is a device made of a weight (called a *bob*) hung from a fixed point so as to swing freely back and forth under the action of gravity. In this unit, students will explore the motion of simple pendulums and compare that motion to that of playground swings.

Science owes much of its understanding about pendulum motion to one of the fathers of the study of physics, Galileo. Legend has it, that when he was 18 years old, Galileo

noticed something unusual about large lamps he observed while sitting in the cathedral of Pisa, Italy. These lamps hung from the ceiling and would often swing back and forth when moved by the lamplighter or a gust of wind. It seemed to the young student that lamps hung the same distance from the ceiling seemed to move in the same rhythm. Galileo used his own pulse to time the swings of the lamps and discovered that the time a pendulum takes to make one complete swing (there and back again) is always the same, even when the pendulum has lost energy and the swings aren't nearly so large. In further experiments, Galileo discovered that different-length pendulums swing in predictable rhythms. Longer pendulums have lower frequency (less swings per minute). Shorter pendulums have higher frequency (more swings per minute). He discovered that he could change the length of a pendulum until it swung exactly 60 times in a minute. This discovery led to the development of pendulum clocks, often referred to as grandfather clocks. See the article "Galileo and the Lamps" (*www.mainlesson.com/display.php?author= baldwin&book=thirty&story=galileo*) for more information about Galileo's discovery.

As students work through the pendulum lab, they will use the same size bob on pendulums of various lengths. Their data will consist of the mass of their bob, the length of the pendulum, and the number of swings they count in each of four minutes from the time the bob is dropped, until it nearly comes to rest four minutes later. Casual observation of the pendulum makes it clear that each swing is a little smaller than the one before. This is because air resistance converts a portion of the pendulum's kinetic energy to heat, thereby reducing the energy available to make the climb on the upswing. A common misconception is that there will be fewer swings in minute #2 than in minute #1 because the pendulum seems to be going so much slower. By making a line plot, students should see the surprising fact that their counts are very consistent, at least over the first three minutes. It is possible that the pendulum could come to rest in minute #4 or shortly thereafter since the washers used as a bob have fairly high air resistance. When the class data are combined in a large line plot, students are likely to be surprised that the other teams who used bobs with less or more mass (each team has a bob with a different mass: one, two, three or four washers) counted swing counts similar to theirs on pendulums with the same length. This is the "aha" we are looking for: While the pendulum keeps swinging, swing count per minute is consistent for pendulums of the same length where the bobs have equivalent air resistance.

In classic pendulum studies, the length of the pendulum is defined as the distance from the top of the string to the center of gravity of the bob. For this lab, the center of gravity is near the center of the washers. You may determine that measuring to the center of the washers is too difficult for your students. In that case, simply have students measure from top of the string to the bottom of the washers when the pendulum is hanging at rest.

Jigsaw Research Method

The jigsaw method is a way for students to become experts on a particular research topic that is part of a larger study. Each student can become an expert for their team by researching a topic with a member of each of the other design teams. While researching, the student will complete the graphic organizer provided, and the group will compile their findings on poster board or chart paper. This poster will serve as a visual with which they will share their knowledge with the class. Suggested topics for this lesson are below.

Research Topic 1: Motion. Applying pendulum principles to swing design requires students to solve an important problem: how to keep the swing going when the natural behavior of the pendulum is to slow and eventually stop. Students will need to devise a propulsion system to fulfill one of the design requirements: Design a swing that adds to the fun factor score for the proposed playground. Each team should appoint a propulsion expert. This student should participate with peers from other teams in the propulsion portion of a jigsaw learning team.

Viewing a trapeze workshop video (such as the one at *https://vimeo.com/143478380*) gives students a clue that thrusting a rider's weight in the direction of the movement of the swing will add additional energy to the swing. Repeated application of additional energy can propel the rider higher and higher. This method of propulsion makes sense for a trapeze rider, but is not practical for most swing configurations.

On a conventional swing, riders add energy by rapidly extending their lower legs, a motion sometimes called pumping. A rider can add even more energy by leaning back and forth so much that the chains (or ropes) that hold the swing actually go from straight to bent with the bend happening where the hands grasp the chains. This minor deflection of the chain lifts the body upward against gravity. This lift, added repeatedly at the correct part of the swing, increases the speed of the swing since the body is falling from a slightly higher location. The "How Swings Work" video from the University of Nottingham (*www.youtube.com/watch?v=UXo6WvHRs_I*) illustrates the idea perfectly.

A wheelchair swing (see an example video at *www.youtube.com/watch?v=ZOzEJrdv1V4*) offers a completely different propulsion alternative. The rider adds energy to the swing by pulling on a stationary rope. Notice that this fulfills the classic definition of force as a push or a pull.

Research Topic 2: Safety. As with any object used by or in contact with living things, safety is an important concern. In designing a swing, your budding engineers will need to consider the safety of the riders as well as the safety of bystanders. Each team should appoint a safety expert. This student should participate with peers from other teams in the safety portion of a jigsaw learning team. The definitive resource for public playground safety guidance is the Public Playground Safety Handbook (*www.cpsc.gov/PageFiles/122149/325.pdf*) published by the U.S. Consumer Product Safety Commission.

Research Topic 3: Materials. One important design consideration is the materials that will be used for the structure. Most children have experience with a wide variety of playground equipment and will be familiar with many kinds of material (e.g., wood, metal, plastic, concrete). Each team should appoint a materials expert. This student should participate with peers from other teams in the materials portion of a jigsaw learning team. The internet has numerous commercial resources for playground materials. One example is a directory of suppliers with links to websites where students can get ideas for their proposals (see *www.playgroundprofessionals.com/equipment/playground/independent-play*).

Research Topic 4: Aesthetics. The swing unit is part of the larger Swing Set Makeover Design Challenge. Some students may see this grand structure as a canvas for artistic expression. School mascots, colors, or design motifs may figure prominently in aesthetic considerations. Each team should appoint an aesthetics expert. This student should participate with peers from other teams in the aesthetics portion of a jigsaw learning team.

For more information on the jigsaw method, see the online overview available at *www.jigsaw.org/overview*.

Career Connections

As career connections related to this lesson, you may wish to introduce the following:

- *Materials Scientist.* The design challenge for this module has a special focus on materials. When students hear the word *material*, they might think of fabric used to make clothes. When engineers talk about materials, they mean the stuff things are made of more generally. Materials can be plastics, metals, ceramics, or some other substance. Materials scientists are people who study and design materials to solve special problems. Explain to students that if they like science and enjoy working in a laboratory, they might enjoy being a materials scientist. For more information about this career, see *www.sciencebuddies.org/science-engineering-careers/engineering/materials-scientist-and-engineer#whatdotheydo*.

- *Landscape Architect.* This module's design focus is a swing set or playground. The design of the outdoor surroundings of the playground is just as important as the playground itself. Landscape architecture is the design of planned changes to the land, especially in areas where people live, work, and play. A landscape architect might work on many different kinds of projects such as parks, gardens, playgrounds, monuments, or city design. Explain to students that if they enjoy being outdoors and making places beautiful and organized, they might enjoy being a landscape architect. You can also share the story of Cornelia Oberlander, who knew she wanted to make parks when she was only 11 years old (see *http://alumni.harvard.edu/stories/cornelia-oberlander*).

Differentiation Strategies

Pendulums. The University of Colorado has an online pendulum simulator located at *http://phet.colorado.edu/en/simulation/pendulum-lab.* This simulator allows the user to explore pendulum motion by varying any or all of four factors: gravity, friction (air resistance), mass of the bob, and length of the pendulum. A few different trials illustrate that the mass of the bob does not affect the time of the swings. On Earth, the main drivers of variation to the time of a single swing are the length of the pendulum and the friction encountered by the bob as it moves through the air. One limitation of the simulation as an extension for this lesson's lab is that the mass cannot be set lower than 0.1 kg. This is the equivalent of about six washers—more than any of the groups will use. Aside from this limitation, the simulations offer great tools.

Graphic Organizers. You can help your students organize and communicate their research more effectively by using graphic organizers. Many graphic organizers are available to compare and contrast content and to determine relationships. The following websites offer various options in your search for an ideal design for student organization:

- *www.teach-nology.com/worksheets/graphic*

- *https://www.eduplace.com/graphicorganizer/*

- *www.internet4classrooms.com/grade_level_help/organize_information_language_arts_third_3rd_grade.htm*

Templates. Templates can help students organize their thinking while asking specific questions or providing space for students to process their reading and thinking. The Lesson 2 handout Writing the OREO Way (p. 138) and the Jigsaw Research Template at the end of this lesson (p. 175) are two examples of how templates can be used to help students of all abilities organize their research to guide thinking. A number of examples are also available online, including the following:

- *www.teachervision.com/writing-research-report-gr-3*

- *www.pinterest.com/pin/469429961130290879*

COMMON MISCONCEPTIONS

Students will have various types of prior knowledge about the concepts introduced in this lesson. Table 4.12 (p. 154) outlines some common misconceptions students may have concerning these concepts. Because of the breadth of students' experiences, it is not possible to anticipate every misconception that students may bring as they approach this lesson. Incorrect or inaccurate prior understanding of concepts can influence student learning in the future, however, so it is important to be alert to misconceptions such as those presented in the table.

Table 4.12. Common Misconceptions About the Concepts in Lesson 3

Topic	Student Misconception	Explanation
Pendulums	A heavier pendulum bob or a lighter pendulum bob will swing faster (shorter period).	The primary determinants of the time it takes a pendulum to go back and forth (the period) are the length of the string and the air resistance on the bob; the weight of the bob is not a factor.
	As a pendulum swings, the shorter swings take less time than the longer swings.	Using his pulse as a timer, Galileo proved that the time a pendulum takes to make one complete swing is always the same.

PREPARATION FOR LESSON 3

Review the Teacher Background Information provided (p. 149), assemble materials for the lesson, and preview videos included within the Learning Components section. Students will be assessed on their team swing unit designs and their presentations of the proposed designs. You may wish to review the associated rubrics (attached at the end of this lesson).

The science centerpiece of this lesson is the pendulum lab. You may wish to conduct this lab on your own prior to working with students since proper setup of the apparatus is essential to its function. You should ensure that there is plenty of room for each team's pendulum to swing without hitting team members. The longest string is approximately 30 inches.

Set up the four pendulum stations as follows:

1. Tape a pencil to the table with the eraser end hanging off the front of the table by several centimeters.

2. Make a loop in one end of the string to provide a place to attach the bob to the string.

3. Fasten one end of the string to the end of the pencil so that the bob can swing freely. Leave enough room to hang the bob below the loop at the end of the string. The goal here is to have several different stations each with a different length of string. If tables are short, the range of the string lengths will need to be adjusted accordingly.

4. Place a line plot recording sheet at each station. Students will use them in the first station, and then carry the graph with them to each station they are able to visit. Each station will have a different color marker for recording their results. The empty line plot should include the following:

- A vertical axis scale from 0 to 100 in increments of 5.

- A horizontal axis that will be reserved for identifying each trial at each length of string.

5. Make a pendulum bob to use to demonstrate the procedure for marking a location for the drop and the proper way to drop the bob (see the Activity/Exploration section on p. 157 for details).

Each design team member will be performing a job in the pendulum lab. Decide in advance how team members will be assigned to jobs (i.e., will teams decide which job each member will do or will you assign jobs). The following jobs are available:

- Technician—holds the bob and drops the bob to start the pendulum.

- Statistician—counts the number of full swings (there and back again) in a specific period of time.

- Quality Control—does the measuring and makes sure the tests are fair. This student will make sure the bob begins at the same point and the record times are consistent. This student will also count the number of full swings of the pendulum.

- Crew Chief—runs the stopwatch, calls the start, and calls out each minute.

LEARNING COMPONENTS

Introductory Activity/Engagement

Connection to the Challenge: Begin each day of this lesson by directing students' attention to the driving question for the module and challenge: How can I use what I know about force and motion to create a plan and build a model of a swing set that is both fun and safe? Discuss what students learned in Lessons 1 and 2 that will be helpful in addressing the challenge and review the driving question for Lesson 3. Have students record their ideas in their STEM Research Notebooks.

Driving Question for Lesson 3: How can I use what I know about force and motion to create a plan and build a model of a swing unit that is both fun and safe?

Science Class: Watch videos that depict different kinds of swings. Remind students that a change in motion is always the result of a push or a pull force. Encourage them to look for pushes and pulls in videos depicting swings such as the following:

- *www.youtube.com/watch?v=ZOzEJrdv1V4*

- *https://vimeo.com/143478380*

- *www.youtube.com/watch?v=UXo6WvHRs_I*

STEM Research Notebook Prompt

Discuss the videos and allow students an opportunity to add to the KWL chart in their STEM Research Notebooks. Use the following questions and prompts to help stimulate ideas:

- *What were the people in the videos doing?*

- *Create a T-chart in your STEM Research Notebook to compare and contrast the swings in the video to those on your playground. Consider the shapes of the swings, their size, and where they are placed (location).*

- *How did the rider make each of the swings go higher?*

- *How did the rider add energy in the path of each swing?*

Mathematics Connection: Not applicable.

ELA Connection: Read aloud the following poem:

> *The Swing*
> By Robert Louis Stevenson
>
> How do you like to go up in a swing,
> Up in the air so blue?
> Oh, I do think it the pleasantest thing
> Ever a child can do!
>
> Up in the air and over the wall,
> Till I can see so wide,
> Rivers and trees and cattle and all
> Over the countryside—
>
> Till I look down on the garden green,
> Down on the roof so brown—
> Up in the air I go flying again,
> Up in the air and down!
>
> *Source:* Public domain.

After reading this poem, ask students to identify words in the poem that describe the motion of swinging. Emphasize the phrases "up in the air and over the wall" and "up in the air and down." Ask students if these are good descriptions of swinging. After a brief discussion, ask students to use their own words to write a single sentence in their STEM Research Notebooks to describe their own experiences with swings.

Read *Follow Me* by Tricia Tusa. This story allows students to experience a girl's imagination as she swings and does a good job of showing students the leg pumping action that is necessary for moving a swing. This may aid your students as they imagine imitating pumping a swing higher without actually moving their feet.

Hold a class discussion about swinging, allowing students to share their own ideas about and experiences with swinging. You may wish to use the following questions and prompts:

- How did you feel as you listened to this story?

- What words in the story helped you experience swinging in your mind?

- Close your eyes and imagine you are swinging. What is your body doing? What are your arms doing? What are your feet doing?

- Consider the forces in your motion. Describe what forces are working on your body.

- Explain how the design of the swing helps or hinders these forces.

Give students an opportunity to add specific details about swings to the KWL chart in their STEM Research Notebooks.

Social Studies Connection: Not applicable.

Activity/Exploration

Science Class: Introduce the pendulum investigations with a class discussion about pendulums using the following questions and prompts:

- *Pendulum* is a science term used to describe a device that includes weight, called a *bob*, that is hung from a point so it can swing freely back and forth by gravity. What part of the playground works like a pendulum?

- When a pendulum begins swinging, the path is long and curved like an arc. The pendulum will swing for a long time—back and forth. If you don't touch the pendulum, will the path of the swing get longer, shorter, or stay the same? Explain your thinking.

- As the path of the pendulum's swing get shorter, does it take less time, more time, or the same amount of time? Explain your thinking.

Pendulum Investigations

Students will explore different length pendulums at each station in these activities. Each group has a set of washers that they will use for their pendulum weight at each station.

The goal is to observe the behavior of each pendulum and compare that behavior to the pendulums with different string lengths. Each team will have a different number of washers for their weight, but they will not know that the other teams' weights are different. During this activity, the design teams will observe that, without the addition of more energy, a pendulum will gradually slow down and its swings get shorter. After making this observation, students will propose ways to add energy to a swing ride on the swing set.

Demonstrate the pendulum set-up and motion to the class as follows:

1. Show students how to set the bob along the edge of the table even with the pencil. Stretch the string until it is tight and lined up along the edge of the desk.

2. Use a piece of tape to mark the place on the desk where the washer is lying. This will be the drop point mark. Explain that to create a fair experiment, this drop point needs to be the same for every test.

3. Show students to proper way to begin the swing, by sliding the washer gently off the edge of the desk and letting it swing freely.

Once you are sure students know how to safely begin the swing, pass out the pendulum investigation handouts and a different number of washers to each design team. Group 1 will receive one washer, Group 2 will receive two washers, Group 3 will receive three washers, and Group 4 will receive four. Modify the number of total washers if necessary.

Before beginning Pendulum Activity #1, student teams will each build a pendulum bob from metal washers using the directions on the Pendulum Investigations handout attached at the end of this lesson (p. 168). Tell students that the washer pack is called the pendulum *bob*. Students should weigh their bobs on a scale and record the weight in grams in their STEM Research Notebooks.

Pendulum Investigation #1: Identify the Path

Ask students, How can we create a plan and build a model for a swing that meets the criteria for fun and safety? You may wish to hold a class discussion to add to the class list of key ideas on chart paper, or have students add to a Module Connections page in their STEM Research Notebooks. Encourage the students to "see the circle" in the pendulum and to transfer that same idea to swings on a typical swing set. This will help them identify the footprint of the swing path to identify the safety zone needed to provide clearance for a swing.

Pendulum Investigation #2: Pendulum Motion

In Pendulum Investigation #2, students rotate through four different stations. Each station has a pendulum set up with a different length string (30 cm, 45 cm, 60 cm, and 75 cm). The string length is the only variable that changes from station to station since each design team will be using a consistent bob weight throughout their trials. Students will measure the number of full swings (there and back again) made during specific intervals: for example, minute 1, minute 2. Pendulum Activity #2 helps students see the effect of forces acting on the pendulum bob, including the following:

- Gravity—pulls the bob down

- Inertia—keeps the bob moving

- Friction—in the form of air resistance, slows it down

- Pull of the string—opposes gravity so the pendulum can keep repeating its cycle because of inertia

Review each type of force with students. Students should be familiar with gravity. Ask students to share their ideas about gravity, guiding students to understand that gravity is a force that pulls two objects together; gravity pulls on people and holds them to the Earth and causes objects to fall (demonstrate by dropping an object on the floor). Remind students of their discussion of friction from Lesson 1. Ask students whether a pendulum swinging in the air experiences friction. Students may be surprised to learn that even though the pendulum does not touch a solid object when it swings that it experiences friction from the air. This is called air resistance. Students may be less familiar with inertia. Introduce inertia as the force that keeps moving objects moving and that keeps resting objects at rest until the objects are acted on by another force. Ask students if they have ever been in a car when it stopped suddenly. Ask them what happened to their bodies when the car stopped (their bodies jerked forward). Tell them that this is an example of inertia. Their bodies kept moving forward until their seat belts stopped them. Ask students if they think a pendulum in motion will keep swinging or if it will stop eventually. If it will stop, what forces are acting on the pendulum that would make it stop swinging (gravity and friction or air resistance).

Draw students' attention to places in the pendulum swing where some or all of the forces are balanced. Ensure that the students see that forces working together and in opposition result in different motions of the pendulum bob:

- At rest, all of the forces are balanced—no movement.

- As the pendulum falls, gravity and inertia work together, opposed only by friction, and the pendulum picks up speed.

- As the pendulum rises, inertia is opposed by gravity and friction working together to slow the pendulum until it reverses at the top of one side or the other.

- Friction is at work on both sides of the swing, dissipating energy from the system.

- The force of the string pulling against gravity enables inertia's effect to repeat over and over again.

Encourage students to predict what they think will happen when they work with pendulums of different lengths:

- Does a longer pendulum mean it swings more times before coming to rest?

- Does a longer pendulum mean more swings per minute?

- How do the number of swings in minute 1 compare with minute 2 and the others?

Have each team share their data from Pendulum Activity #2 on a class table that includes a column for the weight of the bobs and hold a class discussion about whether the weights of the bobs affected the number of pendulum swings. Give students an opportunity to write reflections about the pendulum activities using the STEM Research Notebook prompt that follows. These activities should have resulted in one or two "aha" discoveries (see Explanation section, p. 164). Encourage students to use scientific language and offer details of what they learned, including surprising results (e.g., the weight of the bob did not affect the number of swings).

STEM Research Notebook Prompt

What did you learn about pendulums that surprised you from the pendulum activities? How can you use what you learned to design a swing for your Swing Set Makeover Challenge?

Swing Makeover Plan Book

Introduce the swing makeover portion of the Swing Set Makeover Design Challenge with a class discussion. Ask students for their ideas about how the pendulums they made are the same and different from a swing (e.g., swings hang from a fixed point like a pendulum, swings swing like a pendulum when they are lifted and released, a swing differs from a pendulum when the swing rider begins with a small back-and-forth motion and adds energy). Ask students to respond to the following question: How do riders add energy to a swing to make it go higher?

The swing makeover is the engineering portion of the lesson. Up until now, students have been concerned with swinging motion as illustrated in a simple pendulum. The design challenge requires that students add a new dimension—propulsion. The trapeze and wheelchair videos in the Introductory Activity/Engagement section illustrated two very different propulsion systems: thrusting the body to add momentum and pulling on a rope to move the swing. In addition to these methods, students are probably familiar with the motion of pumping their legs to add momentum. In this section, students should be encouraged to imagine a swing design that incorporates any or all of these propulsion methods, or something entirely different. The design needs to fit within the size constraints of the playground measurements that were identified in the Lesson 2 Geometry Scavenger Hunt.

Students should use their imaginations and creativity rather than focusing on solving mechanical problems in this challenge. The output is a set of design documents showing the primary features of the proposed solution. Each team should produce the following:

- An annotated drawing

- A proposal that identifies the advantages of the team's design

- A materials list

Have students share their opinions on how the pendulums they observed in the activities would score on the Fun Factor Survey from Lesson 1. Here are some questions they could consider:

- If this pendulum was a ride at the playground, what would be its fun factor rating?

- What variable (mass or pendulum length) has the most effect on the fun factor score?

- How could you improve the pendulum ride's fun factor rating?

Tell students that they will work in teams to use the EDP to create a plan for a new swing design. Student teams should work together to work through the steps of the EDP; however, each student should create his or her own "plan book" (see the Swing Makeover Plan Book handout, p. 176) and annotated sketch. After students complete their plan books and sketches, teams should choose one sketch to use for their swing design. This may be one team member's sketch, or teams may choose to create a new sketch that incorporates elements of sketches from more than one team member.

After teams have selected one sketch for their swing designs, have teams review one another's sketches and provide feedback, using the following format:

- I like the following about this swing design

- I think this swing design could be improved by

- I have questions about

Have teams reconvene and discuss the feedback they received. Teams should revise their swing design sketches based on that feedback.

Mathematics Connection: In the pendulum investigations, students will use line plots:

- At each station, each team should measure the length of the pendulum at the station.

- At each station, teams plot the swing counts on a team line plot with columns from 10 to 60 in increments of 5. They should use the colored marker from the station to distinguish counts from the different stations.

- At the end of the activity, students should transfer the team swing counts to a class line plot with columns labeled from 5 to 100. Each team should distinguish their measurements by noting the number of washers in their bob in a circle with color corresponding to the station color.

- Students should compare the number of swings in minutes 1 to 4 for each station (differing pendulum lengths) and each team (differing pendulum weights).

For the Swing Makeover Plan Book activity, use measurements from Pendulum Investigation #1 to model the size of the safety zone required for the swing to keep bystanders safe.

Introduce the concept of scaled drawings. Show students a sheet of graph paper. Remind students that they are designing a model of a swing set rather than a full-size swing set. Ask them if their models will be bigger than the sheet of paper. Refer students to their slide sketches from Lesson 2 where they marked the height and width of their slides. Ask students if the drawing is the size of an actual playground slide (it is not). Tell students that engineers and map makers often have to draw large things on small pieces of paper. They do this by creating scaled drawings. A scaled drawing is a drawing in which the actual sizes of objects are reduced (or enlarged) by a certain amount. Ask students if they had any problems in representing their slide designs accurately in their drawings and if they think that the proportions of their models will be the same as what they have drawn (probably not). Point out to students that using a scale allows designers to create a more accurate picture of what the object will look like when it is completed. Show students a road map. Ask them if the map shows the actual sizes of roads. Tell students that the map makers use a scale. Point out the scale in the map key (e.g., 1 inch = 10 miles). Ask students for ideas about how they could create a scaled drawing using graph paper (e.g., 1 square = 1 meter).

Next, have student teams look back at their slide design sketches. Based on the dimensions of their slide, what size should their swing be (e.g., it should probably be taller than the slide).

ELA Connection: Encourage students to draw conclusions based on their observations. Students should write a reflection in their STEM Research Notebooks about pendulum behavior they observed in the pendulum investigations. You may wish to use the following questions and prompts to guide students' writing:

- How do the swing counts across minutes 1 through 3 vary for the same station at different weights?

- How do the swing counts across minutes 1 through 3 vary for the same weight at different stations?

- Describe the forces acting on the pendulum.

Jigsaw Teamwork Method for Research

Teams will research topics related to playground equipment using a jigsaw strategy. The Teacher Background Information section (p. 149) contains information about jigsaw grouping and about the research topics. Students will need chart paper and markers for this activity.

Before beginning the jigsaw activity, lead a class discussion about swing set safety. Ask students to write five playground rules and have them share these rules with their teams. After students share their rules with their teams, ask them to underline the rules that are rules about having fun. Make a class list of these rules. Now ask students to place a star beside the rules that are safety rules. Ask students whether there are more rules concerned with safety or fun. Ask students whether companies that build swing sets have a responsibility to make their products safe.

As a class, consider some features of playground equipment that are safety features and make a list. Have students decide on several safety features to use as safety criteria for their swing set makeover designs.

In addition to applying pendulum principles to the design of their swings, students will be tasked with four important considerations which will serve as constraints to their designs: propulsion, safety, aesthetics, and materials. To tackle these important problems, employ the jigsaw method of grouping students. The jigsaw method involves dividing students into small groups (called expert groups) to investigate one factor of a larger phenomenon. Students then present these findings to the larger group so that, together, the class forms an overview of the various factors, using the presentations to put them together, much like a puzzle is constructed. For this lesson, each expert group will be charged with investigating one factor of playground design. You should ensure that each design team has a member on each of the expert teams, so that the design team has an "expert" on each factor represented; the number of expert groups should match the number of members on each design team as closely as possible.

The expert groups should research applicable standards and safety guidelines for playgrounds. The Teacher Background Information section (p. 149) will help you better understand the groups' research topics.

Social Studies Connection: Encourage students to look at the swing set from a mechanical engineer's point of view. Examine the materials list the students created in the slide and swing design plan book activities. Identify safety standards that might be applied to the materials and construction of the slide and swing to keep children safe and make the playground long-lasting. Have students write a paragraph reflection in their STEM Research Notebooks about why companies that manufacture playground equipment are responsible for providing safe, good-quality products.

Explanation

Science Class: Pendulum Investigation #1 helps students recognize that the path of an object with a fixed center point and fixed distance (radius) between the object and the center point is a circle. In the real world, a pendulum follows an arc, which is a part of the circle, having its center point at the point of the circle closest to the Earth. Once students start to see this circular pattern for motion, they will see circles everywhere (e.g., the motion of a swinging arm, the path of bicycle pedals, the arc of a door). The pendulum's path is an arc at the bottom of the circle. As the energy of the pendulum is diminished by friction, the arc gets smaller and smaller.

Mathematics Connection: Pendulum Investigation #1 sets the baseline for measurements in Pendulum Investigation #2. Each team gets a different number of washers. The pendulum data provides an opportunity for student "aha" moments. The data will show that longer pendulum lengths cause fewer swings per minute. The various bob masses should have little or no effect on the swing counts. Students should see that to provide a fast swing ride with more oscillations, a long pendulum length is not needed.

The swing makeover portion of the Swing Set Makeover Design Challenge provides an opportunity to calculate areas and perimeters using the size of the proposed swing design, called the *footprint*. In mechanical engineering, an object's footprint is the size and shape of the surface that the object covers. Guide the teams as they project the swing's total horizontal movement down onto the ground. This is the footprint and should be enlarged to establish a safety zone for bystanders. Guide the students into determining how big of a buffer is needed around the swing's footprint—possibly an arm's length.

ELA and Social Studies Connections: The findings from the jigsaw research and the output from science and mathematics components of the swing makeover sections should be combined in a proposal for the team's swing design concept. Have each student team create a two-to-five paragraph write-up of the proposal for the team's swing design. Teams should use the Proposal Presentation and Swing Set Makeover rubrics (located at the end of Lesson 4) to ensure they include the appropriate information in their proposals.

Elaboration/Application of Knowledge

Science Class: Once students complete the pendulum investigations, they may have a new awareness of and appreciation for pendulum devices, such as grandfather clocks and yo-yos. Encourage students to look at their bodies and try to find a pendulum (their arm is a pendulum—it swings back and forth from a fixed point in a straight line). A number of animals also have pendulum limbs. A horse's legs, for example, are two pairs of pendulums that swing in opposite directions.

Mathematics Connection: This is an excellent time to introduce the concept of strength of materials. The string used in the pendulum lab is adequate to hold a few washers, but would not be sufficient to hold something heavy like a bowling ball or a person. Similarly, the tape on the desk is adequate to secure the pencil used to suspend the pendulum above the ground. Neither the tape nor the pencil would adequately support anything very heavy, however. Have student teams brainstorm ideas for materials for their model swings and materials for actual swings that people would use. Students can enter these ideas in a T-chart in their STEM Research Notebooks.

ELA and Social Studies Connections: The jigsaw research activity provides an opportunity to consider the responsibilities of citizens to work together for the common good. In this case, citizens could be companies that manufacture equipment used in playgrounds and schools. Students should be encouraged to see that companies and individuals have a civic responsibility for public safety and health. This responsibility should go beyond the repercussions in the marketplace of selling poor-quality or unsafe items. Companies should provide safe, high-quality products because it is the right thing to do.

Evaluation/Assessment

Students may be assessed on the following performance tasks and other measures listed.

Performance Tasks

- Pendulum investigations
- Swing Makeover Plan Book and sketch
- Jigsaw research
- Written proposal for swing design

Other Measures

- Engagement in class activities and discussions
- Involvement in group work and discussions

- STEM Research Notebook entries
- Collaboration (see rubric in Lesson 2, p. 131)

INTERNET RESOURCES

"Galileo and the Lamps" article

- *www.mainlesson.com/display.php?author=baldwin&book=thirty&story=galileo*

Trapeze workshop video

- *https://vimeo.com/143478380*

"How Swings Work" video from the University of Nottingham

- *www.youtube.com/watch?v=UX06WvHRs_I*

Video about a wheelchair swing

- *www.youtube.com/watch?v=ZOzEJrdv1V4*

Public Playground Safety Handbook

- *www.cpsc.gov/PageFiles/122149/325.pdf*

Information about playground materials

- *www.playgroundprofessionals.com/equipment/playground/independent-play*

Information about jigsaw learning teams

- *www.jigsaw.org/overview*

Information about materials scientists

- *www.sciencebuddies.org/science-engineering-careers/engineering/materials-scientist-and-engineer#whatdotheydo*

Article about a landscape architect (Cornelia Oberlander)

- *http://alumni.harvard.edu/stories/cornelia-oberlander*

Online pendulum simulator

- *http://phet.colorado.edu/en/simulation/pendulum-lab*

Information about graphic organizers

- *www.teach-nology.com/worksheets/graphic*

- *https://www.eduplace.com/graphicorganizer/*

- *www.internet4classrooms.com/grade_level_help/organize_information_language_arts_third_3rd_grade.htm*

Information about templates

- *www.teachervision.com/writing-research-report-gr-3*

- *www.pinterest.com/pin/469429961130290879*

Name: _____   Team Name: _____

PENDULUM INVESTIGATIONS

SAFETY NOTES

1. All laboratory occupants must wear safety glasses or goggles during all phases of this inquiry activity.

2. Make sure there are no fragile materials in the area where activities are taking place.

3. Have an appropriate level of adult supervision to ensure safe behaviors during activities.

4. Make sure all materials are put away after completing the activity.

5. Wash hands with soap and water after completing this activity.

INSTRUCTIONS FOR SET UP

1. Assemble a washer pack by stacking the washers your teacher gives you together and wrapping a pipe cleaner around the washers by passing the pipe cleaner through the holes in the center of the washers several times. Twist the pipe cleaner together to hold the washers tightly. Leave enough of the pipe cleaner to fasten to the string of the pendulum. This group of washers is called the pendulum bob. Your design team will use this bob for the entire activity.

2. Identify which team members have the following jobs:

   - Technician—holds the bob and drops the bob to start the pendulum.

   - Statistician—counts the number of full swings (there and back again) in a specific period of time.

   - Quality Control—does the measuring and makes sure the tests are fair. This student will make sure the bob begins at the same point and the record times are consistent. This student will also count the number of full swings of the pendulum.

   - Crew Chief—runs the stopwatch, calls the start, and calls out each minute.

3. Measure and record the weight of the washers (in grams) using a scale. What do they weigh? Record your answer in your STEM Research Notebook.

Name: _____   Team Name: _____

PENDULUM INVESTIGATION #1: IDENTIFY THE PATH

PROCEDURE

1. The technician should fasten the bob to the loop at the end of the string by twisting the loose ends of pipe cleaner together through the loop.

2. The quality control student should mark your drop point with masking tape (set the bob along the edge of the table even with the pencil, stretch the string until it is tight and lined up along the edge of the desk. Use a piece of tape to mark the place on the desk where the washer is lying. This will be the drop point mark.

3. The statistician should measure the length of the string (in cm) from the pencil to the bottom of the bob. Record the length in your STEM Research Notebook.

4. Draw a sketch of the pendulum (with the string length and bob weight labeled) in your STEM Research Notebook.

5. The technician should gently move the bob to the drop point and drop the bob, making sure the pendulum's string is tight through the entire movement.

6. The crew chief should hold a meter stick upright with the end on the floor near the path of the pendulum but far enough away that the pendulum will not hit it.

7. The crew chief should give the signal for the bob to be dropped. Make sure the string remains tight through the entire movement.

8. The crew chief should call out the height the bob reaches on the meter stick at the end of the first full swing and then at the end of the third full swing, at the end of the fifth swing, and then again at the end of the seventh swing of the pendulum. The statistician should write these measurements down:

   Bob height at end of first swing: _____ cm

   Bob height at end of third swing: _____ cm

   Bob height at end of fifth swing: _____ cm

   Bob height at end of seventh swing: _____ cm

Swing Set Makeover Lesson Plans

Name: _____   Team Name: _____

PENDULUM INVESTIGATION #1: IDENTIFY THE PATH

9. All team members should watch to identify the path of the bob. What shape is being formed?

10. Add to the pendulum sketch you made in your STEM Research Notebook showing the pendulum's path. Label locations on the path, including the drop point, the highest point, and the lowest point.

Name: _____   Team Name: _____

PENDULUM INVESTIGATION #1: IDENTIFY THE PATH

INVESTIGATION QUESTIONS

1. How is a pendulum's motion like a swing's motion?

2. What happened to the height of the bob as the pendulum continued to swing?

3. How can you use information about the pendulum's swing in your swing design?

4. How can you use information about the pendulum's swing to be sure that your swing design is safe?

Name: _____   Team Name: _____

PENDULUM INVESTIGATION #2: PENDULUM MOTION

PROCEDURE

1. Review the data table for this investigation where you will record your results (provided below). You will be recording the following:
   - The length of the string at each of four stations
   - The number of swings of the pendulum for each minute of a three-minute trial
   - The height the bob reaches at the end of each minute

2. The technician should fasten the bob to the loop at the end of the string by twisting the loose ends of pipe cleaner together through the loop.

3. The quality control student should mark your drop point with masking tape. To do this, set the bob along the edge of the table even with the pencil, stretch the string until it is tight and lined up along the edge of the desk. Use a piece of tape to mark the place on the desk where the washer is lying. This will be the drop point mark.

4. The statistician should measure the length of the string (in cm) from the pencil to the bottom of the bob. Record the length in your STEM Research Notebook.

5. Place the pendulum bob at the drop mark. The string should be straight and tight along the edge of the table.

6. The crew chief should say "go" and start the stopwatch at the same time. The technician should release the bob when the crew chief says "go."

7. The statistician should hold a meter stick upright with the end on the floor near the path of the pendulum but far enough away that the pendulum will not hit it.

8. All team members should observe the pendulum for three minutes. The quality control student should count the number of full swings (there and back is one) in one minute.

9. After one minute has passed, the crew chief should call "one" and the technician should write down the number of swings in that minute.

10. At one minute, the statistician should note the height of the pendulum and record that.

11. The quality control student should start counting swings again (beginning at 1) after the crew chief signals that one minute has passed.

12. The crew chief should say "two" after two minutes have passed, and the technician should record the number of swings in minute two.

13. At two minutes, the statistician should note the height of the pendulum and record that.

14. The quality control student should start counting swings again (beginning at 1) after the crew chief signals that two minutes have passed.

Name: _____   Team Name: _____

PENDULUM INVESTIGATION #2: PENDULUM MOTION

15. The crew chief should say "three" after three minutes have passed, and the technician should record the number of swings in minute three.

16. At three minutes, the statistician should note the height of the pendulum and record that.

17. The statistician should add the three numbers for swings per minute together to get the total for the number of swings in three minutes.

18. The technician should remove the team's bob from the string before going to the next station.

19. Record the total scores on the line plot paper at each station using the marker that is there. Each length of string will have a different color marker to make it easier for you to keep track of your data.

20. Repeat Steps 3–19 at each station.

DATA TABLE FOR PENDULUM INVESTIGATION #2

String Length	Swings in Minute 1	Swings in Minute 2	Swings in Minute 3	Total Swings in 3 Minutes	Bob Height at End of Minute 1	Bob Height at End of Minute 2	Bob Height at End of Minute 3
30 cm							
45 cm							
60 cm							
75 cm							

Name: _____   Team Name: _____

PENDULUM INVESTIGATION #2: PENDULUM MOTION

INVESTIGATION QUESTIONS

1. What gives the pendulum its energy?

2. How does a swing get energy to keep it moving?

3. How could we add more energy to the pendulum?

4. What are other ways to add energy to a swing?

Name: _____

STUDENT HANDOUT

JIGSAW RESEARCH TEMPLATE

Topic _____

Your group sent you to get ideas about a problem they are trying to solve. Get together with other experts for this topic and learn things that only experts know!

Main Idea #1
Main Idea #2
Main Idea #3
Resources

Name: _____   Team Name: _____

Our swing set name is _____

STUDENT HANDOUT, PAGE 1

SWING MAKEOVER PLAN BOOK

Use the following tables to help you work through your team's swing design like an engineer. You will need four pieces of paper. Put the pages together and fold them in half, like a book. Give it a title, then turn the page and begin. You will have two pages of your book for each of the first three tasks. You will create your drawing (the fourth task) on graph paper and then glue or staple it into your book.

TASK 1: DEFINE THE PROBLEM—FIND THE IMPORTANT PARTS

Description	Fun Factor Strong Points
What is the role of a swing on the swing set?	What fun factor elements are met by a swing's role?
Safety	**Other Things to Consider**
What safety rules apply to swings?	What other limits might be important?

Name: _____   Team Name: _____

Our swing set name is _____

SWING MAKEOVER PLAN BOOK

TASK 2: IMAGINE IT!—BRAINSTORM POSSIBLE SOLUTIONS

Ideas to Add Function How can you change the swing to add new ways to play on it?	**Ideas for Fun** What can you do to the swing to make it more fun?
Safety Pointers How could you make the swing even safer?	**More Ideas to Improve the Swing** What other ideas do you have to improve your swing?

TASK 3: PLAN IT!—RESEARCH, LIST MATERIALS, AND IDENTIFY THE NEXT STEPS

Functions: Research It! How have others changed the design of a swing? (Cite your resources.)	**Fun: Research It!** How have others made the swing more fun? (Cite your resources.)
Safety: Research It! What are the rules for safety that every swing must have?	**More Ideas: Research It!** What else did you learn when you were doing your research?

Swing Set Makeover Lesson Plans

Name: _____   Team Name: _____

Our swing set name is _____

SWING MAKEOVER PLAN BOOK

TASK 4: SKETCH IT!—FOLLOW YOUR PLAN AND DRAW YOUR SWING DESIGN

After filling out the information above, make a sketch with your ideas on graph paper. Label the parts of the swing, the sizes of each part, and the materials you will use. Add notes for anything else that will help make your design easier to understand when you use it to build your swing. After you complete your sketch, glue or staple it into your plan book.

Be sure to include the following:

- Labels for the center point of the swing arc, the preferred height of the swing above ground, and the width of the swing seat

- Footprint of the swing arc marked with a dotted line to indicate the clearance required for safety

- Notes indicating the propulsion system for the swing unit

- A materials list

TASK 5: SHARE IT!—ENGINEERS SHARE THEIR IDEAS AND USE FEEDBACK

Share your sketch with your team members. Compare all of the team members' sketches and discuss what you like about each one. Choose one sketch to use as your team's swing design, or create a new sketch using things you like from more than one sketch.

Pendulum Investigations Rubric

Name: _____

Criteria	Needs Improvement (1 point)	Approaching Standard (2 points)	Meets Standard (3 points)	Exceeds Standard (4 points)	Score
DATA AND RESULTS	• Much of the data are not recorded and it is not possible to find a pattern. • Data show evidence that the team did not follow directions.	• Data are recorded, but it is not possible to make clear predictions. • Data show evidence that the team often did not follow directions.	• Data are recorded so that patterns can be seen. • Data show evidence that the team mostly followed instructions.	• Data are recorded neatly so that patterns can be seen. • Data show evidence that the team followed all instructions.	
DISCUSSION	• Provides short answers. • Concepts seem to be unclear and mixed. • No connections are made to the real world. • Uses no key terms.	• Includes some detail in answers. • Understanding of concepts is shaky. • Connections made to the real world are confusing. • Key terms are not used correctly.	• Includes detailed answers. • Understands most concepts. • Makes connections to the real world with prodding. • Uses some key terms.	• Includes detailed answers. • Shows strong evidence of understanding. • Makes clear connections to the real world. • Correctly uses science key terms.	
COMPLETE	• Several required elements are missing.	• One element is missing, but additional parts have been added that make the results clearer.	• All the required elements are present in the final report.	• All the parts are present and additional elements have been added to make the results clearer.	

TOTAL SCORE: _____

COMMENTS:

Swing Design Sketch Rubric

Name: _____   Team Name: _____

Points	Standards	Details
25	• Detailed notes describe the parts of the swing (e.g., measurements, materials). • The swing design is practical (i.e., if the design were used, a working swing would result). • The swing design clearly reflects understanding of pendulum motion. • Reasoning for material choices is provided and clearly explained. • The design is unique and creative.	• The sketch is neat and may be colored with colored pencils. • A key for scale is provided.
20	• Notes help describe the parts of the swing. • The swing design is practical (i.e., if the design were used, a working swing would result). • The swing design reflects understanding of pendulum motion. • Some reasoning for material choices is provided. • The design is different from the examples shown.	• The sketch is somewhat neat and may be colored with crayons. • A key for scale is provided.
10	• Some notes are provided, but are not very helpful. • It is not clear that the swing design is practical (i.e., if the design were used, a working swing may or may not result). • The swing design reflects limited understanding of pendulum motion. • Reasoning for material choices is provided but is unclear. • The design is similar to the examples shown.	• The sketch is not as neat as it could be and does not include color. • A key for scale is provided but may be unclear or difficult to understand.
5	• Sketch does not have notes to help explain the features of the swing. • The swing design is impractical (i.e., if the design were used, a working swing would not result). • The swing design reflects little or no understanding of pendulum motion. • No reasoning for materials choices is provided. • The design is a copy of the examples shown.	• The sketch is not neat and is not colored. • A key for scale is not provided.

SCORE: _____

COMMENTS:

Lesson Plan 4: Swing Set Makeover Design Challenge

This lesson serves as the capstone of the module and challenges students to complete their swing set makeovers. The science and mathematics activities guide student teams through building a model of the proposed designs. Students use the EDP to take their sketches to physical form. In ELA, students document this process and discuss the reasoning for their design choices and the benefits of their designs in a classroom blog. In social studies, students extend their discussion about parks from previous lessons by focusing on state and national parks.

ESSENTIAL QUESTIONS

- How can we use what we know about forces to design a playground swing set?

- What safety features need to be considered when designing swing sets?

- What arguments can be used to prove to others that the swing set designed by my team is a good design?

- How can citizens help state and national parks?

ESTABLISHED GOALS AND OBJECTIVES

At the conclusion of this lesson, students will be able to do the following:

- Build a model for a new swing set that considers shapes and forces and has a high rating on the Fun Factor Survey from Lesson 1

- Build a model for a new swing set that incorporates an understanding of basic safety features

- Effectively use shapes, materials, and measurements that affect speed, aesthetics, and safety on a new swing set design

- Communicate information about swing set designs in a presentation

- Create a blog or newsletter that addresses the design decisions made in the design process and the benefits of the swing set design

- Recognize that citizens have civic responsibilities, which include helping care for public parks

- List ways that they, as citizens, can help care for public parks

TIME REQUIRED

- 8–10 days (approximately 45 minutes each day; see Tables 3.9–3.10, pp. 42–43)

MATERIALS

Handouts for Lesson 4

- EDP Cards handout (1 per student)
- Measurement Stations handout (1 per student)
- Swing Set Makeover Planning handout (1 per student)
- Peer Review Form (1 per student)
- Swing Set Makeover Graphic Organizer (1 per student)
- Building a Model handout (1 per team)

Rubrics for Lesson 4

- Team Blog Rubric
- Proposal Presentation Rubric
- Swing Set Makeover Rubric
- Collaboration Rubric (see Lesson 2, p. 131)

Necessary Materials for Lesson 4

- STEM Research Notebook
- Internet access

Additional Materials for Science Class (Model Building)

- Measuring and building devices (per team)
  - Ruler or meter stick
  - Scissors
  - Needle nose pliers
  - Assortment of tools to work with modeling clay
  - Camera or other device for taking photographs

- Possible building materials include the following:

  - Modeling clay

  - Straws

  - Pipe cleaners

  - Cardboard (or cardstock paper)

  - Note cards

- Possible fastening materials include the following:

  - String

  - Tape

  - Craft glue or tacky glue

  - Wire

  - Glue gun and hot glue

  - Indirectly vented chemical splash goggles

Additional Materials for Mathematics Connection

- Measuring stations (1 per design team) with various items to measure, such as the following:

  - Regularly shaped objects, such as a card, box, and book

  - Irregularly shaped objects, such as a ball, bowl, action figure, and toy

- 12-inch ruler (4 per station)

- Scale (1 per station)

- String (2 18-inch lengths per station)

Additional Materials for ELA and Social Studies Connections

- The book *Up, Up, in a Balloon* by Lawrence F. Lowery (part of the *I Wonder Why* series, NSTA Kids, 2013)

- Writing paper and pen

- 11-inch × 17-inch construction paper or copy paper (1 per student)

SAFETY NOTES

1. All laboratory occupants must wear safety glasses or goggles during all phases of this inquiry activity.

2. Make sure there are no fragile materials in the area where activities are taking place.

3. Have an appropriate level of adult supervision to ensure safe behavior during activities.

4. Use caution when working with heat-producing equipment (e.g., glue gun and hot glue) because they can severely burn skin.

5. Use caution when working with sharps (e.g., scissors, pliers, wires, pipe cleaners) to avoid cutting or puncturing skin or eyes.

6. Only use GFI-protected electrical receptacles for power sources.

7. Make sure all materials are put away after completing the activity.

8. Wash hands with soap and water after completing this activity.

CONTENT STANDARDS AND KEY VOCABULARY

Table 4.13 lists the content standards from the *NGSS, CCSS,* and the Framework for 21st Century Learning that this lesson addresses, and Table 4.14 (p. 189) presents the key vocabulary. Vocabulary terms are provided for both teacher and student use. Teachers may choose to introduce some or all of the terms to students.

Table 4.13. Content Standards Addressed in STEM Road Map Module Lesson 4

NEXT GENERATION SCIENCE STANDARDS

PERFORMANCE EXPECTATIONS

- 3-PS2-1. Plan and conduct an investigation to provide evidence of the effects of balanced and unbalanced forces on the motion of an object.

- 3-PS2-2. Make observations and/or measurements of an object's motion to provide evidence that a pattern can be used to predict future motion.

- 3-5-ETS1-1. Define a simple design problem reflecting a need or a want that includes specified criteria for success and constraints on materials, time, or cost.

- 3-5-ETS1-2. Generate and compare multiple possible solutions to a problem based on how well each is likely to meet the criteria and constraints of the problem.

- 3-5-ETS1-3. Plan and carry out fair tests in which variables are controlled and failure points are considered to identify aspects of a model or prototype that can be improved.

Continued

Table 4.13. (*continued*)

SCIENCE AND ENGINEERING PRACTICES

Asking Questions and Defining Problems

- Ask questions about what would happen if a variable is changed.

- Ask questions that can be investigated and predict reasonable outcomes based on patterns such as cause and effect relationships.

- Use prior knowledge to describe problems that can be solved.

- Define a simple design problem that can be solved through the development of an object, tool, process, or system and includes several criteria for success and constraints on materials, time, or cost.

Developing and Using Models

- Collaboratively develop and/or revise a model based on evidence that shows the relationships among variables for frequent and regular occurring events.

- Develop a model using an analogy, example, or abstract representation to describe a scientific principle or design solution.

- Develop a diagram or simple physical prototype to convey a proposed object, tool, or process.

- Use a model to test cause and effect relationships or interactions concerning the functioning of a natural or designed system.

Planning and Carrying Out Investigations

- Plan and conduct an investigation collaboratively to produce data to serve as the basis for evidence, using fair tests in which variables are controlled and the number of trials considered.

- Evaluate appropriate methods and/or tools for collecting data.

- Make observations and/or measurements to produce data to serve as the basis for evidence for an explanation of a phenomenon or test a design solution.

Using Mathematics and Computational Thinking

- Decide if qualitative or quantitative data are best to determine whether a proposed object or tool meets criteria for success.

- Describe, measure, estimate, and/or graph quantities (e.g., area, volume, weight, time) to address scientific and engineering questions and problems.

Constructing Explanations and Designing Solutions

- Use evidence (e.g., measurements, observations, patterns) to construct or support an explanation or design a solution to a problem.

- Identify the evidence that supports particular points in an explanation.

- Apply scientific ideas to solve design problems.

- Generate and compare multiple solutions to a problem based on how well they meet the criteria and constraints of the design solution.

Continued

Table 4.13. (*continued*)

> *Engaging in Argument From Evidence*
> - Compare and refine arguments based on an evaluation of the evidence presented.
> - Distinguish among facts, reasoned judgment based on research findings, and speculation in an explanation.
> - Respectfully provide and receive critiques from peers about a proposed procedure, explanation, or model by citing relevant evidence and posing specific questions.
> - Construct and/or support an argument with evidence, data, and/or a model.
> - Use data to evaluate claims about cause and effect.
> - Make a claim about the merit of a solution to a problem by citing relevant evidence about how it meets the criteria and constraints of the problem.
>
> *Obtaining, Evaluating, and Communicating Information*
> - Read and comprehend grade-appropriate complex texts and/or other reliable media to summarize and obtain scientific and technical ideas and describe how they are supported by evidence.
> - Compare and/or combine across complex texts and/or other reliable media to support the engagement in other scientific and/or engineering practices.
> - Combine information in written text with that contained in corresponding tables, diagrams, and/or charts to support the engagement in other scientific and/or engineering practices.
> - Obtain and combine information from books and/or other reliable media to explain phenomena or solutions to a design problem.
> - Communicate scientific and/or technical information orally and/or in written formats, including various forms of media as well as tables, diagrams, and charts.
>
> ## DISCIPLINARY CORE IDEAS
>
> *PS2.A: Forces and Motion*
> - Each force acts on one particular object and has both strength and a direction. An object at rest typically has multiple forces acting on it, but they add to give zero net force on the object. Forces that do not sum to zero can cause changes in the object's speed or direction of motion.
> - The patterns of an object's motion in various situations can be observed and measured; when that past motion exhibits a regular pattern, future motion can be predicted from it.
>
> *PS2.B: Types of Interactions*
> - Objects in contact exert forces on each other.
>
> ## CROSSCUTTING CONCEPTS
>
> *Patterns*
> - Patterns of change can be used to make predictions.
> - Patterns can be used as evidence to support an explanation.

Continued

Table 4.13. (*continued*)

Structure and Function

- Different materials have different substructures, which can sometimes be observed.

- Substructures have shapes and parts that serve functions.

Influence of Science, Engineering, and Technology on Society and the Natural World

- People's needs and wants change over time, as do their demands for new and improved technologies.

- Engineers improve existing technologies or develop new ones to increase their benefits, to decrease known risks, and to meet societal demands.

COMMON CORE STATE STANDARDS FOR MATHEMATICS

MATHEMATICAL PRACTICES

- MP1. Make sense of problems and persevere in solving them.

- MP2. Reason abstractly and quantitatively.

- MP4. Model with mathematics.

- MP7. Look for and make use of structure.

MATHEMATICAL CONTENT

- 3.MD.C.5. Recognize area as an attribute of plane figures and understand concepts of area measurement.

- 3.MD.C.6. Measure areas by counting unit squares (square cm, square m, square inches, square feet, and improvised units).

- 3.MD.C.7. Relate area to the operations of multiplication and addition.

- 3.OA.D.9. Identify arithmetic problems (including patterns in the addition table or multiplication table), and explain them using properties of operations.

COMMON CORE STATE STANDARDS FOR ENGLISH LANGUAGE ARTS

READING STANDARDS

- RI.3.5. Use text features and search tools (e.g., key words, sidebars, hyperlinks) to locate information relevant to a given topic efficiently.

- RI.3.7. Use information gained from illustrations (e.g., maps, photographs) and the words in a text to demonstrate understanding of the text (e.g., where, when, why, and how key events occur).

- RI.3.8. Describe the logical connection between particular sentences and paragraphs in a text (e.g., comparison, cause/effect, first/second/third in a sentence).

- RI.3.10. By the end of the year, read and comprehend informational texts, including history/social studies, science, and technical texts, at the high end of the grades 2–3 text complexity band independently and proficiently.

Continued

Table 4.13. (*continued*)

WRITING STANDARDS

- W.3.1. Write opinion pieces on topics or texts, supporting a point of view with reasons.

- W.3.1.A. Introduce the topic or text they are writing about, state an opinion, and create an organizational structure that lists reasons.

- W.3.1.B. Provide reasons that support the opinion.

- W.3.1.C. Use linking words and phrases (e.g., because, therefore, since, for example) to connect opinion and reasons.

- W.3.1.D. Provide a concluding statement or section.

- W.3.2. Write informative/explanatory texts to examine a topic and covey ideas and information clearly.

- W.3.2.B. Develop the topic with facts, definitions, and details.

- W.3.3. Write narratives to develop real or imagined experiences or events using effective technique, descriptive details, and clear event sequences.

SPEAKING AND LISTENING STANDARDS

- SL.3.1. Engage effectively in a range of collaborative discussions (one-on-one, in groups, and teacher-led) with diverse partners on *grade 3 topics and texts,* building on others' ideas and expressing their own clearly.

- SL.3.1.D. Explain their ideas and understanding in light of the discussion.

- SL.3.3. Ask and answer questions about information from a speaker, offering appropriate elaboration and detail.

- SL.3.4. Report on a topic or text, tell a story, or recount an experience with appropriate facts and relevant, descriptive details, speaking clearly at an understandable pace.

- SL.3.6. Speak in complete sentences when appropriate to task and situation in order to provide requested detail or clarification.

FRAMEWORK FOR 21ST CENTURY LEARNING

- Interdisciplinary Themes: Health and Safety; Environmental Literacy; Science; Mathematics; Engineering Design Process (EDP)

- Learning and Innovation Skills: Creativity and Innovation; Critical Thinking and Problem Solving; Communication and Collaboration

- Information, Media, and Technology Skills: Information Literacy; Media Literacy; ICT Literacy

- Life and Career Skills: Flexibility and Adaptability; Initiative and Self-Direction; Social and Cross-Cultural Skills; Productivity and Accountability; Leadership and Responsibility

Table 4.14. Key Vocabulary for Lesson 4

Key Vocabulary	Definition
blog	a web journal; a space for sharing ideas and writing about what you know in an informal format online

TEACHER BACKGROUND INFORMATION

The teacher support material for taking the design sketches from Lessons 2 and 3 and converting them into a model has been provided as a handout attached at the end of the lesson (Building a Model, p. 205). You may wish to copy this for student use.

Blogs

Student teams will create web logs, or blogs, about their swing set designs and design decisions. Blogs can be great educational tools, giving students freedom to publish content on the web. Blog writing, unlike typical academic writing, is informal, and therefore may be less intimidating to students. The first step for blog creation is choosing a platform on which to build and publish the blogs. There are many secure sites where you can do this for free. The following are examples of educational blog sites:

- *Edublogs.org,* which lets you create and manage teacher and student sites. This site is popular because it is highly customizable. You can make tweaks to change colors and designs, and add photos, videos, and even podcasts.

- *Kidblog.org* is a blog platform designed for grades K–12. It is free for up to 50 students per class. Some of the features include privacy, password protection, simplicity, and no advertising.

- *WordPress.org* is another option, although the public has access to blogs created here. It is not as simple to use as those specifically designed for educators, but it has some great features and includes a number of "plug-ins" especially designed to make it versatile.

For more information about classroom blogging, see the following:

- *www.educationworld.com/a_tech/tech/tech217.shtml*

- *www.edutopia.org/blog/blogging-in-21st-century-classroom-michelle-lampinen*

- *www.edutopia.org/blog/introducing-social-media-lower-elementary-beth-holland*

- *http://edublogs.org/10-ways-to-use-your-edublog-to-teach*

If online publishing is not available, students may create a newsletter in lieu of a blog.

You may wish to have each team member be responsible for one component of the blog or newsletter. Examples of components to assign include the following:

- A documentary of the process in which the model was developed—a personalized walk-through of the EDP as it applied to the team's own design creation and model building, which would highlight the reasons for the team's design decisions

- An advertisement for their swing set targeted to the school decision makers who are responsible for buying swing set equipment—this component would highlight the benefits of the team's design

- An article about the safety features of the swing set

- A science activity, different from one those done during the module (e.g., a "You Can Do This at Home" feature article), which should address one of the science concepts from the module (e.g., gravity, friction, unbalanced forces, ramps, pendulums).

COMMON MISCONCEPTIONS

Students will have various types of prior knowledge about the concepts introduced in this lesson. Table 4.15 outlines some common misconceptions students may have concerning these concepts. Because of the breadth of students' experiences, it is not possible to anticipate every misconception that students may bring as they approach this lesson. Incorrect or inaccurate prior understanding of concepts can influence student learning in the future, however, so it is important to be alert to misconceptions such as those presented in the table.

Table 4.15. Common Misconceptions About the Concepts in Lesson 4

Topic	Student Misconception	Explanation
Models	Models are art projects.	Models are used to demonstrate and explain concepts that may be difficult to describe using only words. They may contain artistic elements, but their purpose is more than artistic.
	Models need to show every part of the object they represent.	Models should show the major features of the object they represent but do not need to include every detail.
	Models cannot be changed once they are constructed.	Using the EDP to construct a model means that the model can and should be changed and improved on so that it does a better job demonstrating and explaining the function of the object it represents.

PREPARATION FOR LESSON 4

Review the Teacher Background Information provided (p. 189), assemble materials for the lesson, and preview videos included within the Learning Components section. Students will be assessed on their collaboration, their teams' models, and their contributions to their teams' blog posts. You may wish to review the associated rubrics (attached at the end of this lesson).

Prepare measurement stations with objects for students to measure. Prepare the same number of stations as there are design teams in the class. Each design team will visit three stations. Include some objects that are easy to measure (e.g., a card, a box, a book) and some objects that are more difficult to measure (e.g., a ball, a bowl, an irregularly shaped item such as an action figure or toy).

Students will apply the learning they have accumulated throughout the module in this lesson. You may wish to refresh your memory of the EDP before reading the book *Up, Up in a Balloon* by Lawrence F. Lowery with students and read the book ahead of time. You should begin assembling the materials for the swing set models before teams begin building. At this point in the module, students should have completed their swing set and slide sketches and excitement should be growing for building their models.

Before you begin the blog project in ELA, check with your school administrator for guidelines for student blogging and use this to develop your goals and expectations. For example, some schools may not allow photos of students on blog posts or may require written permission from parents to participate in a blog assignment.

At the end of the lesson, you will have guests (e.g., school administrators, older students) act as potential "customers" to give feedback on teams' models. Issue invitations to guests.

LEARNING COMPONENTS
Introductory Activity/Engagement

Connection to the Challenge: Begin each day of this lesson by directing students' attention to the driving question for the module and challenge: How can I use what I know about force and motion to create a plan and build a model of a swing set that is both fun and safe? Review the driving question for Lesson 4 and discuss what students learned in the previous lessons that will be helpful in addressing the challenge. Have students record their thinking in their STEM Research Notebooks.

Driving Question for Lesson 4: How can I use what I have learned in the module to create a plan and build a model of a swing set that is both fun and safe?

Science Class: Remind students that they will be using the designs they created in the previous lessons to build their models. Remind students about the steps of the EDP and tell students that people who create new products (engineers and inventors) often have to move back to previous steps of the EDP in the building phase. Ask students for their ideas about why this might be (e.g., they find problems in their plan when they begin to build and need to adjust drawings). Engage the class in a discussion using the following questions:

- What kind of person makes a good inventor?

- If an inventor's first idea for solving a problem fails, does that mean that the problem cannot be solved?

- What are some good ways to get new ideas to solve a problem?

EDP Cards Activity

Distribute an EDP Cards handout to each student. Each handout consists of seven cards, one for each step of the design process. Review the words on each card and guide students in recalling their meaning and application in the EDP. Read a selection from *Up, Up in a Balloon* or another book that illustrates the EDP. As you read, encourage students to hold up their card when they perceive an element of the EDP in action.

Remind students that the EDP is used to solve problems that may seem like impossible tasks, like the problem the Montgolfier brothers faced in *Up, Up in a Balloon*. Let students share their own ideas or stories about times when they tried something at which they were not successful on the first try (e.g., riding a bicycle, swimming, or playing a game). Have students respond to the following STEM Research Notebook prompt and then hold a class discussion about students' ideas and experiences about failure and perseverance in achieving goals, relating these experiences to the EDP and how design teams might respond to difficulties or failure.

STEM Research Notebook Prompt

Name a time when you tried something new and were not successful on the first try but tried again. Answer the following questions:

- *How did you feel when you failed?*

- *Who or what encouraged you to try again?*

- *How did it feel to finally succeed at something that was difficult?*

Mathematics Connection: Remind students that in Lesson 1 they developed a way to measure the fun factor of a piece of playground equipment. Ask students to describe how they measured the fun factor. Hold a class discussion about whether that kind of measurement takes the place of measurements like length, width, height, and mass by using the following questions:

- Can a measurement like the fun factor measurement take the place of engineering measurements like length, width, height, and mass? Why or why not?

- Are engineering measurements like length, width, height, and mass meaningful for things like a book or a video or a painting?

- How do you decide what measurements are meaningful?

ELA Connection: Introduce the concept of a blog as an online journal or way to share ideas with others. Ask students to share their experiences with journaling or blogs. Provide some examples of blogs created by elementary school students (see sites referenced in the Teacher Background Information section, p. 189). As a class, discuss features of the blogs, using questions and prompts such as the following:

- What do you notice about how the blogs look? Are there pictures?

- What is the purpose of the blog? (e.g., to convey information, to tell a story, to share an opinion)

- How is the writing style similar to or different from the writing in your textbooks?

- How is the writing style similar to or different from a letter you would write to a friend?

- If you have kept a journal or diary, how is the blog similar to or different from your journal or diary?

Social Studies Connection: Not applicable.

Activity/Exploration

Science Class: Review the Swing Set Makeover Design Challenge with the class, emphasizing that the goal is to create a model of a swing set that will meet their standards for fun (as defined by their Fun Factor Surveys) while incorporating safety considerations. After teams' models are developed, students will share their designs and defend their reasoning for design decisions in a blog.

In previous lessons, each design team created a sketch for a slide design and a sketch for a swing design. Tell students that in this lesson their teams will convert these designs into physical models. Hold a class discussion about the definition of the term *model*, guiding students to an understanding that a model is a representation of an object showing its major features and is usually a smaller-scale version of the original.

Begin the lesson with a class discussion reminding students about models in Lesson 1. Discussion prompts may include the following:

- Have you ever used a model to make something? What were you making?

- How is a model helpful in the designing stage of a project?

- Does a model need to be able to do everything that the full-size object can do? Explain your answer.

Each team should next devise a way to combine their designs for their slide and their swing into a single playground unit. This may require an additional design (for example a connecting climbing unit or tower). Refer to the Building a Model handout attached at the end of the lesson (p. 205) for instructions about the workflow to produce a model.

Share with the class the following guidelines for their models:

- The swing set (swing, slide, and connector) should be designed so that it will fit in the site determined in Lesson 2.

- The model should be built on a piece of stiff board, such as a 15-inch × 20-inch foam display board.

- The model need not be fully functional but should include the primary features of the full-size swing set.

- It is all right if, when building the model, the teams need to vary the original designs for their swing or slides. Emphasize that solving design problems is one reason for building a model.

- If the swing or slide designs are modified in the building stage, the original sketches should be revised to match the model.

- A design sketch similar to the one prepared for the swing and the slide should be prepared for the connecting unit to join the swing and slide units.

Measurement Stations

Mathematics Connection: As the students build their swing set models, they will need to take measurements of the materials they are using to determine how they will fit in the overall swing set. Introduce the idea that measurement involves three steps:

1. Identify the features of the object that can be measured.

2. Determine what unit (e.g., inches, square feet, meters, pounds) is best for measuring the feature.

3. Determine what tool should be used to make the measurement.

As a class, work through the three steps to measure and weigh two or three items of varying shapes, demonstrating how to use a string to measure objects with rounded surfaces (e.g., wrap the string around the surface to be measured, marking the beginning and ending points of the measurement with your fingers and having a student use a ruler to measure the length of the string).

Have each student team cycle through three measurement stations (see the Preparation for Lesson 4 section, p. 191). Each student should choose one item to measure at each station and determine what features can be measured for the item (the Measurement Stations handout, p. 200, has space for three objects). Students should determine the best way to measure the measurable features and note those measurements on their copy of the handout. Hold a class discussion about the objects that were easy to measure and objects that were difficult to measure. Have students share features of items that they measured and those they did not measure, commenting on why those chose to measure or not to measure features.

Blog or Newsletter Project

ELA Connection: Lead a discussion with the class using the following prompts:

- Imagine that your school is going to purchase one of the swing sets that are being developed by the teams in your class. What information would they need to make a decision about which swing set to purchase?

- What information would a swing set company want to tell customers about their swing sets?

Student teams will share their swing set designs and defend their reasoning for their design decisions in a blog (if an online blog platform is not available, the students can develop a series of newsletter articles meeting the same requirements). Blogs should serve as a platform for convincing the public of the superiority of the team's design. The blogs also allow teams to view other teams' designs. Teams should include images of their designs in their blogs. See the Teacher Background Information section (p. 189) for ideas about how to divide work among team members.

Social Studies Connection: Extend the discussion about parks from previous lessons to focus on state and national parks. The National Park Service's Find a National Park Service Map web page, located at *www.nps.gov/hfc/cfm/carto.cfm*, illustrates the distribution of parks across the United States. Comparing and contrasting the number and sizes of parks in various states can provide a geography connection for students.

Discuss with students that citizens support state and national parks with tax dollars as well as by paying entrance fees (this is different from the city parks). Hold a class discussion about visitors' responsibilities to maintain the park (e.g., not littering, not taking anything out of the park).

An optional extension is for students to create placemats for a local restaurant with games and information focusing on city, state, and national parks as a culminating project for this module's social studies theme of civic responsibility. Games and information could address park safety, care of parks, park features, and park jobs.

Explanation

Science Class: Building the model may be daunting to the students. The fact that their sketches lack detail or precision shouldn't be viewed as a hindrance. The objects they have designed exist in their imagination in surprising detail. The Building a Model handout (attached at the end of this lesson on p. 205) can help add structure to the process of turning their vision into a three-dimensional model. Encourage students to understand that a model need not have all the details of real life.

Mathematics Connection: Review measurement units with students, highlighting the difference between the English (or U.S. customary) and metric systems of measurement.

Have students share everyday examples of each of these measurements (e.g., speed limit signs in miles per hour, soda containers in liters). Create a T-chart to record students' ideas. Hold a class discussion about which measurement system is most frequently used.

ELA Connection: Have students create presentations for their swing sets. These presentations should incorporate the information in their blogs. The Proposal Presentation Rubric (attached at the end of the lesson on p. 210) contains the criteria for presentations.

Social Studies Connection: Not applicable.

Elaboration/Application of Knowledge

Science Class: Have each team present its swing set design and design process to the class and guests (e.g., school administrators, older students). The audience should act as potential "customers" for the teams' swing sets and give feedback on the models and the presentations. Provide guidelines to visitors for feedback, such as the ones that follow:

- Things I like about your swing set
- Things that could be improved about your swing set
- Questions I have about your swing set

Mathematics Connection: Not applicable.

ELA Connection: Have student teams review each other's blog posts. A Peer Review Form is provided at the end of this lesson (p. 202).

Social Studies Connection: Not applicable.

Evaluation/Assessment

Students may be assessed on the following performance tasks and other measures listed.

Performance Tasks

- Model for the Swing Set Makeover Design Challenge
- Blog that documents team use of the EDP and well as the benefits of the proposed design
- Peer Review Form for evaluating the blog or newsletter

Other Measures

- Engagement in class activities and discussions
- Involvement in group work and discussions

- STEM Research Notebook entries
- Rubrics provided for collaboration (see Lesson 2), presentation, and writing assignments

INTERNET RESOURCES

Blog websites
- *Edublogs.org*
- *Kidblog.org*
- *WordPress.org*

Articles and resources about blogging
- *www.educationworld.com/a_tech/tech/tech217.shtml*
- *www.edutopia.org/blog/blogging-in-21st-century-classroom-michelle-lampinen*
- *www.edutopia.org/blog/introducing-social-media-lower-elementary-beth-holland*
- *http://edublogs.org/10-ways-to-use-your-edublog-to-teach*

Find a National Park Service Map web page
- *www.nps.gov/hfc/cfm/carto.cfm*

EDP CARDS

DEFINE

Identify the problem.

LEARN

Brainstorm possibilities.
Choose the best one.

PLAN

Research. List needs.
Plan your steps.

TRY

Follow your plan.
Build a model.

TEST

Try it out! What works?
What doesn't work?

DECIDE

Redesign the model to solve problems
that came up.

SHARE

Show it to others to get feedback.

When you see or hear a step in
the EDP being used in the story,
hold up your card.

Swing Set Makeover Lesson Plans

Name: _____

Team Name: _____

MEASUREMENT STATIONS

What **FEATURES** can I measure?	What **UNITS** fit the features?	What **TOOLS** measure those units?
• Length • Color • Width • Thickness • Mass • Fun Factor • Other?	• Inches • Meters • Miles • Pounds • Grams • Liters • Other?	• Ruler • Microscope • Scale • Beaker • Tape Measure • Other?

ANSWER EACH MEASURING QUESTION FOR ONE OBJECT IN EACH STATION

What is the object?	What features can I measure?	What units fit the features?	What tools measure those units?
What is the object?	What features can I measure?	What units fit the features?	What tools measure those units?
What is the object?	What features can I measure?	What units fit the features?	What tools measure those units?

NATIONAL SCIENCE TEACHERS ASSOCIATION

Name: _____   Team Name: _____

STUDENT HANDOUT

SWING SET MAKEOVER PLANNING

Step 1: (Define)   State the problem. What are you trying to do?

Step 2: (Learn)   What solutions can you and your team imagine?

Step 3: (Plan)   What features and functionality will make your swing set fun and safe?

Step 4: (Try)   Do your design documents tell the most important ideas about your proposal?

Step 5: (Test)   How did it work? Could it work better?

Step 6: (Decide)   This is your chance to make changes! What did you change?

Step 7: Share your design! What is your swing set fun factor rating? What design features of your swing set do you want to highlight in your blog? Think about what makes it fun, safe, and appealing to your readers.

Reviewer's Name: _____   Name of Team That Authored Blog: _____

STUDENT HANDOUT, PAGE 1

PEER REVIEW FORM

1. The most interesting part of this blog is

2. The least interesting part of this blog is

Reviewer's Name: _____   Name of Team That Authored Blog: _____

STUDENT HANDOUT, PAGE 2

PEER REVIEW FORM

3. My suggestions for the authors are

4. A question I have for the authors is

5. I would give the following grade for this blog:

___ Fair (1)   ___ Average (2)   ___ Above Average (3)   ___ Excellent (4)

SWING SET MAKEOVER GRAPHIC ORGANIZER

Use these steps to create your plan for the new swing set.

Problem

The school has asked for design proposals for a new swing set. Your team has been challenged to design a section of the equipment. The proposal should be for a unit that is safe and provides high fun factor ratings.

Task 1: Your team's design should achieve high fun factor ratings while being safe. Use this Swing Set Makeover Graphic Organizer to gather information for your design planning.

Task 2: What features and functionality will make yours a winning design? Combine these ideas into a sketch and proposal that will describe what your unit will do and how it will work.

DEFINE: Identify the problem

LEARN: Brainstorm possibilities, choose the best

PLAN: Research, list needs, plan steps

TRY: Follow your plan and build a model

Swing Set Makeover Graphic Organizer

Use these steps to create your plan for the new swing set

TEST: Try it out. What works? What doesn't?

DECIDE: Redesign to solve problems that came up

SHARE: Show it to others and get feedback.

Task 3: Use your sketch and notes to make a model of the swing set you have imagined. Your materials list helped you have everything you need!

Task 4: Show your design documents and model to family members and fellow students. Does it convince them that your ideas for the new swing set would be fun? What suggestions do they have?

Task 5: Did you get any ideas about how you could make your swing set better? Make changes to improve your model. These changes should be shown on your sketch, too!

Task 6: It's done! Share your ideas for a swing set makeover in a blog post!

Name: _____   Team Name: _____

BUILDING A MODEL

OVERVIEW

The design challenge in this lesson is to build a model of your team's concept for a new and improved swing set. The general steps for building this model are listed below. Photograph the project frequently as it comes together. These photos can be used later to illustrate the stages of the project for your blog or newsletter.

PROCEDURE

1. Lay out the playground site on graph paper that is the same size as the construction platform (foam board or cardboard).

2. Plan the arrangement of the swing set sections (slide, swing, climber, or tower) on the playground site.

3. Identify the major parts of each section of the swing set.

4. Choose materials that will be used to build the major parts.

5. Build the major parts of the swing set.

6. Assemble the major parts into the swing set model.

7. Add details and finishing touches to the model.

Your team should approach each of these steps like a mini EDP. Each step has a problem (DEFINE), different possible solutions (LEARN), a plan (PLAN), an implementation (TRY), trial and error (TEST), refinement (DECIDE), and the opportunity to get feedback (SHARE). The seven steps are detailed below.

PROBLEM 1. LAY OUT THE PLAYGROUND SITE

1. Fasten multiple sheets of graph paper together to make a sheet the same size as the construction platform. Make sure the sheets are cut and joined so the lines and spaces match.

Name: _____   Team Name: _____

BUILDING A MODEL

2. Determine the scale as follows:

   - Count a single line of squares across the length of the graph paper. If you are using a 15-inch x 20-inch board, this will be about 80 squares.

   - Count a single line of squares down the width of the graph paper. If you are using a 15-inch x 20-inch board, this will be about 60 squares. Determine the scale to use as follows:

   - If the longest side of the site is over 80 feet long, use 1 square = 2 feet

   - If the longest side of the site is between 40 feet long and 79 feet long, use 1 square = 1 foot

   - If the longest side of the site is between 20 feet long and 39 feet long, use 2 squares = 1 foot

   - If the longest side of the site is between 10 feet long and 19 feet long, use 4 squares = 1 foot

3. Sketch the site on the graph paper.

4. Hot glue or double-stick tape the graph paper site drawing to the construction platform, ensuring that the lines of the graph paper run parallel with the edges of the platform.

Name: _____   Team Name: _____

STUDENT HANDOUT, PAGE 3

BUILDING A MODEL

PROBLEM 2. PLAN THE ARRANGEMENT OF THE SWING SET SECTIONS

1. On separate pieces of graph paper, draw the footprint of each section of the swing set.
   - The lines of the graph paper should run the same way as the main feature of the section.
2. Label and cut out the footprint drawings.
3. Arrange the footprint drawings of the sections on the site layout.
4. Move the footprint drawings around to find the best fit for a high fun factor score.
5. Ensure that there are adequate safety zones around each section. Reposition if necessary.

PROBLEM 3. IDENTIFY THE MAJOR PARTS OF EACH SWING SET SECTION

1. Make a list of the major parts of each section. These might include the following:
   - Swing unit frame, seat, chains
   - Sliding surface, steps, ladder
   - Connector sides, steps
2. On separate pieces of graph paper, draw the major parts.
   - The lines of the graph paper should run the same way as the main feature of the part.

PROBLEM 4. CHOOSE MATERIALS FOR EACH MAJOR PART

Different parts will require different materials:
   - Cardboard or clay can be used for walls and platforms.
   - Cardboard can be used for decks.
   - Straws or craft sticks can be used for beams or bars.
   - String can be used for rope or chains.
   - Pipe cleaners can be used for fittings and fasteners.
   - Glue and tape can be used for fasteners.

Name: _____   Team Name: _____

BUILDING A MODEL

PROBLEM 5. BUILD THE MAJOR PARTS OF THE SWING SET
1. As much as possible, build the different parts separately.
2. Modify the design of major parts if necessary.
3. Build the major parts using the sketches from Problem 3.

PROBLEM 6. ASSEMBLE MAJOR PARTS INTO THE SWING SET
1. Using the arrangement determined in Problem 2, arrange the main sections in place on the construction platform.
2. Once fit is established, fasten the sections in place and assemble the sections.
3. Join the sections to each other if the swing set is intended to be a continuous unit.

PROBLEM 7. ADD DETAILS AND FINISHING TOUCHES
1. Decorate and add details to the swing set structure as desired.
2. Add a project label to the construction platform where it can be clearly seen—usually near the front edge.
   - Project Name
   - Team Name
   - Date
3. Add any other desired finishing touches such as shrubs or ground cover.

Team Blog Rubric

Name: _____

Criteria	Needs Improvement (1 point)	Approaching Standard (2 points)	Meets Standard (3 points)	Exceeds Standard (4 points)	Team Score
CONTENT	Blog does not demonstrate understanding of the topic and provides little interesting information about the topic.	Blog shows limited evidence of understanding and provides some information about the topic.	Blog shows evidence of understanding and provides interesting information about the topic.	Blog shows clear evidence of understanding and provides new and interesting information about the topic.	
	Blog includes little or no information.	Blog presents information, but with few or no supporting details.	Blog presents information with some supporting details.	Blog presents focused and well-supported information.	
	Blog posts do not reflect self-awareness or an acknowledgment of the audience.	Self-awareness is reflected, but understanding of the audience is not reflected in the blog posts.	Self-awareness and some understanding of audience are reflected in the blog posts.	Self-awareness and awareness and understanding of the audience are reflected in the blog posts.	
USE OF GRAPHICS AND MEDIA	Blog does not use any graphics or multimedia.	Blog uses some graphics and multimedia, but they are of low quality or do not enhance the content.	Blog uses some graphics and multimedia to enhance the content.	Blog uses high-quality graphics and multimedia that enhance the content.	
	Blog does not cite any sources nor uses captions to describe content.	Blog includes some captions and citations, but some are missing or incomplete.	Blog includes captions and cites sources for most images and multimedia content.	Blog includes captions and cites sources for all images and multimedia content.	
RESEARCH	Blog does not include quotes or facts.	Blog includes quotes or facts, but they may not enhance understanding of the topic.	Blog includes a few quotes or new facts that enhance understanding of the topic.	Blog includes several quotes and new facts that enhance understanding of the topic.	
PROOFING	Blog has many spelling or grammar errors.	Blog has several spelling or grammar errors.	Blog has a few spelling or grammar errors.	Blog is free of spelling or grammar errors.	

TOTAL SCORE: _____

COMMENTS:

Proposal Presentation Rubric (30 points possible)

Team Name: _____

Criteria	Needs Improvement (1–2 points)	Approaching Standard (3–4 points)	Meets or Exceeds Standard (5–6 points)	Team Score
INFORMATION	• Few details about the swing set are given. • Nothing seems to be special about this swing set.	• Some details about the swing set are provided. • Ways this swing set is special are highlighted.	• Many details about the swing set are provided. • Ways this swing set is special are highlighted.	
THE PLAN	• Information is random and no strategy is evident • No introduction is used. • No conclusion is provided. • Time is wasted due to lack of preparation. • The presentation is very short or long.	• Focus of presentation may wander. • An introduction is used. • A conclusion summary is provided. • Some parts are hard to understand. • The time is used satisfactorily. • The presentation is slightly less than 5 minutes or longer than 7 minutes.	• Information is focused and clear. • The introduction gets audience's attention. • The conclusion summarizes the strengths of the swing set. • The plan is easy to understand. • The time is used well and it is obvious the team has practiced. • The presentation is between 5 and 7 minutes long.	
YOUR STYLE	• Only one or two team members share. • Presenters are difficult to understand. • No science or key terms are used.	• All team members participated. • Volume may be too low or speech too fast to understand easily. • Presenters use science words and identify some scientific observations.	• All team members participated. • Presenters are easy to understand. • Presenters use science words and key terms.	
VISUALS	• Visual aids may be poorly presented or distract from the presentation.	• Visual aids are used but do not add to the presentation.	• Visuals aids or media components are used.	
QUESTIONS	• The team fails to respond to questions from audience or responds inappropriately.	• The team responds appropriately to audience questions, but responses may be brief, incomplete, or unclear.	• The team answers appropriately, clearly, and in detail to audience questions.	

TOTAL SCORE: _____

COMMENTS:

Points Possible	Swing Set Makeover Rubric (42 points possible) Team Name: _____	Score
	DEFINE: Identify the Problem	
5–6	• Design team has a clear understanding of the Swing Set Makeover Design Challenge and its requirements. • Design team can use other ideas or practices and apply them to the challenge/problem. • Design team can predict and solve design problem.	
3–4	• Design team can describe the Swing Set Makeover Design Challenge and its requirements. • Design team can recall other ideas or practices that are similar and apply them to the challenge/problem. • Design team is able to predict and solve design problems that are similar to those encountered in class.	
1–2	• Design team needs help to understand the Swing Set Makeover Design Challenge. • Design team may be able to recall some ideas or practices and apply them if they are exactly the same. • Design team is not able to predict or solve design problems that arise.	
	LEARN: Brainstorm Possibilities and Choose the Best Solutions	
5–6	• All ideas are focused on solving the problem while brainstorming. • Design team considers unlikely solutions/models. • Design team uses key ideas and concepts correctly to construct possible solutions.	
3–4	• Some ideas are focused on solving the problem while brainstorming. • Design team considers multiple solutions/models. • Design team uses the best option that was presented in class.	
1–2	• Few ideas are focused on solving the problem while brainstorming. • Design team requires guidance to generate ideas for solving the problem. • Design team does not provide any original ideas to solve problems.	

Continued

Swing Set Makeover Rubric (*continued*)

PLAN: Research, List Needs, and Plan Steps		
5–6	• Design team critically investigates the problem and uses information from many sources. • Design team provides a complete list of the materials needed. • Design team plans and explains the steps that are used, including every detail.	
3–4	• Design team analyzes and selects information from several recommended sources. • Design team provides a list with most of the materials needed. • Design team plans and explains most of the steps that are used with a few details missing.	
1–2	• Design team investigates the problem and collects little information from other sources. • Design team lists some of the materials needed. • Design team plans and explains some of the steps that are needed.	
TRY: Follow Your Plan and Build a Model		
5–6	• Design team provides drawings with every detail shown to provide a guide to build the model. • Design team provides evidence to support that the design meets or exceeds fun factor and safety criteria. • Design team creates more than one design option and evaluates and justifies the chosen model.	
3–4	• Design team provides drawings with enough information to build the model but details have been left out. • Design team provides evidence to support that the design meets either fun factor or safety criteria. • Design team generates an alternate design option after one design fails to meet criteria.	
1–2	• Design team provides rough sketches of the design that are not helpful in model building. • Design team does not provide evidence to support the design decisions that are made. • Design team is not able to create an alternate design solution when the first design fails.	

Continued

Swing Set Makeover Rubric (*continued*)

TEST: Try It Out! What Works? What Doesn't?		
5–6	• Design team is able to choose appropriate materials to build the swing set model and use them effectively. • Design team follows plans as written and completes every detail as described. • Testing is done to show that the swing set model meets all criteria, and test results are well documented and offer clear conclusions for its success.	
3–4	• Design team is able to choose appropriate materials to build the swing set model and can use most of the materials effectively. • Design team carries out the plan but overlooks one or two steps in the design. • The model is tested and meets some of the criteria, and test results are documented and offer reasonable conclusions.	
1–2	• Design team does not choose appropriate materials to build the swing set model or did not use materials effectively. • Design team does not follow the instructions outlined in the plan. • Design team does not test the swing set model adequately; thus, test results are misleading or not fully developed.	
DECIDE: Redesign to Solve Problems That Come Up		
5–6	• All of the design team's decisions are supported with evidence. • The design team bases success on criteria for safety as well as its standards for fun. • Realistic and creative designs are made based on peer feedback.	
3–4	• Some of the design team's decisions are supported with evidence. • Design team bases successful design on fun factor criteria, but not on the safety standards. • Realistic designs are not very creative, but are based on peer feedback.	
1–2	• Design team's decisions are not supported with evidence. • Design team does not consider safety or the fun factor survey criteria when evaluating the design. • Designs are not practical or realistic.	

Continued

Swing Set Makeover Rubric (*continued*)

SHARE: Show It to Others and Get Feedback		
5–6	• Design team received feedback from peers with a variety of opinions to improve the design. • Design team sought feedback from a parent or teacher to get feedback on design. • Design team evaluated the pros and cons of feedback and made changes to the swing set based on feedback.	
3–4	• Design team received feedback from team members to improve designs. • Design team has not sought feedback from adult experts, parents, or teacher. • Design team considered but did not use feedback to make changes to the swing set design.	
1–2	• Design team received feedback from a team member. • Design team has not sought feedback from any other person. • Design team does not make changes based on feedback.	

TOTAL SCORE: _____

COMMENTS:

TRANSFORMING LEARNING WITH SWING SET MAKEOVER AND THE *STEM ROAD MAP CURRICULUM SERIES*

Carla C. Johnson

This chapter serves as a conclusion to the Swing Set Makeover integrated STEM curriculum module, but it is just the beginning of the transformation of your classroom that is possible through use of the *STEM Road Map Curriculum Series*. In this book, many key resources have been provided to make learning meaningful for your students through integration of science, technology, engineering, and mathematics, as well as social studies and English language arts, into powerful problem- and project-based instruction. First, the Swing Set Makeover curriculum is grounded in the latest theory of learning for students in grade 3 specifically. Second, as your students work through this module, they engage in using the engineering design process (EDP) and build prototypes like engineers and STEM professionals in the real world. Third, students acquire important knowledge and skills grounded in national academic standards in mathematics, English language arts, science, and 21st century skills that will enable their learning to be deeper, retained longer, and applied throughout, illustrating the critical connections within and across disciplines. Finally, authentic formative assessments, including strategies for differentiation and addressing misconceptions, are embedded within the curriculum activities.

The Swing Set Makeover curriculum in The Represented World STEM Road Map theme can be used in single-content classrooms (e.g., science) where there is only one teacher or expanded to include multiple teachers and content areas across classrooms. Through the exploration of the Swing Set Makeover Design Challenge, students engage in a real-world STEM problem on the first day of instruction and gather necessary knowledge and skills along the way in the context of solving the problem.

The other topics in the *STEM Road Map Curriculum Series* are designed in a similar manner, and NSTA Press has additional volumes in this series for this and other grade levels and plans to publish more. The volumes covering Innovation and Progress have been published and are as follows:

- *Amusement Park of the Future, Grade 6*

- *Construction Materials, Grade 11*

- *Harnessing Solar Energy, Grade 4*

- *Transportation in the Future, Grade 3*

- *Wind Energy, Grade 5*

The tentative list of other books includes the following themes and subjects:

- The Represented World

  - Car crashes

  - Changes over time

  - Improving bridge design

  - Packaging design

  - Patterns and the plant world

  - Radioactivity

  - Rainwater analysis

- Cause and Effect

  - Influence of waves

  - Hazards and the changing environment

  - The role of physics in motion

- Sustainable Systems

  - Creating global bonds

  - Composting: Reduce, reuse, recycle

  - Hydropower efficiency

  - System interactions

- Optimizing the Human Experience
  - Genetically modified organisms
  - Mineral resources
  - Rebuilding the natural environment
  - Water conservation: Think global, act local

If you are interested in professional development opportunities focused on the STEM Road Map specifically or integrated STEM or STEM programs and schools overall, contact the lead editor of this project, Dr. Carla C. Johnson (*carlacjohnson@purdue.edu*), associate dean and professor of science education at Purdue University. Someone from the team will be in touch to design a program that will meet your individual, school, or district needs.

APPENDIX

CONTENT STANDARDS ADDRESSED IN THIS MODULE

NEXT GENERATION SCIENCE STANDARDS

Table A1 (p. 220) lists the science and engineering practices, disciplinary core ideas, and crosscutting concepts this module addresses. The supported performance expectations are as follows:

- 3-PS2-1. Plan and conduct an investigation to provide evidence of the effects of balanced and unbalanced forces on the motion of an object.

- 3-PS2-2. Make observations and/or measurements of an object's motion to provide evidence that a pattern can be used to predict future motion.

- 3-5-ETS1-1. Define a simple design problem reflecting a need or a want that includes specified criteria for success and constraints on materials, time, or cost.

- 3-5-ETS1-2. Generate and compare multiple possible solutions to a problem based on how well each is likely to meet the criteria and constraints of the problem.

- 3-5-ETS1-3. Plan and carry out fair tests in which variables are controlled and failure points are considered to identify aspects of a model or prototype that can be improved.

Table A1. *Next Generation Science Standards (NGSS)*

Science and Engineering Practices

ASKING QUESTIONS AND DEFINING PROBLEMS

- Ask questions about what would happen if a variable is changed.

- Identify scientific (testable) and non-scientific (non-testable) questions.

- Ask questions that can be investigated and predict reasonable outcomes based on patterns such as cause and effect relationships.

- Use prior knowledge to describe problems that can be solved.

- Define a simple design problem that can be solved through the development of an object, tool, process, or system and includes several criteria for success and constraints on materials, time, or cost.

DEVELOPING AND USING MODELS

- Collaboratively develop and/or revise a model based on evidence that shows the relationships among variables for frequent and regular occurring events.

- Develop a model using an analogy, example, or abstract representation to describe a scientific principle or design solution.

- Develop a diagram or simple physical prototype to convey a proposed object, tool, or process.

- Use a model to test cause and effect relationships or interactions concerning the functioning of a natural or designed system.

PLANNING AND CARRYING OUT INVESTIGATIONS

- Plan and conduct an investigation collaboratively to produce data to serve as the basis for evidence, using fair tests in which variables are controlled and the number of trials considered.

- Evaluate appropriate methods and/or tools for collecting data.

- Make observations and/or measurements to produce data to serve as the basis for evidence for an explanation of a phenomenon or test a design solution.

- Make predictions about what would happen if a variable changes.

ANALYZING AND INTERPRETING DATA

- Represent data in tables and/or various graphical displays (bar graphs, pictographs and/or pie charts) to reveal patterns that indicate relationships.

- Analyze and interpret data to make sense of phenomena, using logical reasoning, mathematics, and/or computation.

- Compare and contrast data collected by different groups in order to discuss similarities and differences in their findings.

- Analyze data to refine a problem statement or the design of a proposed object, tool, or process.

Continued

Table A1. (*continued*)

USING MATHEMATICS AND COMPUTATIONAL THINKING

- Decide if qualitative or quantitative data are best to determine whether a proposed object or tool meets criteria for success.

- Organize simple data sets to reveal patterns that suggest relationships.

- Describe, measure, estimate, and/or graph quantities (e.g., area, volume, weight, time) to address scientific and engineering questions and problems.

- Create and/or use graphs and/or charts generated from simple algorithms to compare alternative solutions to an engineering problem.

CONSTRUCTING EXPLANATIONS AND DESIGNING SOLUTIONS

- Use evidence (e.g., measurements, observations, patterns) to construct or support an explanation or design a solution to a problem.

- Identify the evidence that supports particular points in an explanation.

- Apply scientific ideas to solve design problems.

- Generate and compare multiple solutions to a problem based on how well they meet the criteria and constraints of the design solution.

ENGAGING IN ARGUMENT FROM EVIDENCE

- Compare and refine arguments based on an evaluation of the evidence presented.

- Distinguish among facts, reasoned judgment based on research findings, and speculation in an explanation.

- Respectfully provide and receive critiques from peers about a proposed procedure, explanation, or model by citing relevant evidence and posing specific questions.

- Construct and/or support an argument with evidence, data, and/or a model.

- Use data to evaluate claims about cause and effect.

- Make a claim about the merit of a solution to a problem by citing relevant evidence about how it meets the criteria and constraints of the problem.

OBTAINING, EVALUATING, AND COMMUNICATING INFORMATION

- Read and comprehend grade-appropriate complex texts and/or other reliable media to summarize and obtain scientific and technical ideas and describe how they are supported by evidence.

- Compare and/or combine across complex texts and/or other reliable media to support the engagement in other scientific and/or engineering practices.

- Combine information in written text with that contained in corresponding tables, diagrams, and/or charts to support the engagement in other scientific and/or engineering practices.

- Obtain and combine information from books and/or other reliable media to explain phenomena or solutions to a design problem.

- Communicate scientific and/or technical information orally and/or in written formats, including various forms of media as well as tables, diagrams, and charts.

Continued

Swing Set Makeover, Grade 3

Table A1. (*continued*)

Disciplinary Core Ideas

PS2.A: FORCES AND MOTION

- Each force acts on one particular object and has both strength and a direction. An object at rest typically has multiple forces acting on it, but they add to give zero net force on the object. Forces that do not sum to zero can cause changes in the object's speed or direction of motion.

- The patterns of an object's motion in various situations can be observed and measured; when that past motion exhibits a regular pattern, future motion can be predicted from it.

PS2.B: TYPES OF INTERACTIONS

- Objects in contact exert forces on each other.

Crosscutting Concepts

CAUSE AND EFFECT

- Cause and effect relationships are routinely identified, tested, and used to explain change.

PATTERNS

- Patterns of change can be used to make predictions.

- Patterns can be used as evidence to support an explanation.

STRUCTURE AND FUNCTION

- Different materials have different substructures, which can sometimes be observed.

- Substructures have shapes and parts that serve functions.

INFLUENCE OF SCIENCE, ENGINEERING, AND TECHNOLOGY ON SOCIETY AND THE NATURAL WORLD

- People's needs and wants change over time, as do their demands for new and improved technologies.

- Engineers improve existing technologies or develop new ones to increase their benefits, to decrease known risks, and to meet societal demands.

Source: NGSS Lead States. 2013. *Next Generation Science Standards: For states, by states.* Washington, DC: National Academies Press. *www.nextgenscience.org/next-generation-science-standards.*

Table A2. Common Core Mathematics and English Language Arts (ELA) Standards

MATHEMATICAL PRACTICES	READING STANDARDS

MATHEMATICAL PRACTICES

- MP1. Make sense of problems and persevere in solving them.
- MP2. Reason abstractly and quantitatively.
- MP4. Model with mathematics.
- MP5. Use appropriate tools strategically.
- MP6. Attend to precision.
- MP7. Look for and make use of structure.

MATHEMATICAL CONTENT

- 3.MD.A.2. Measure and estimate liquid volumes and masses of objects using standard units of grams (g), kilograms (kg), and liters (l). Add, subtract, multiply, or divide to solve one-step word problems involving masses or volumes that are given in the same units, e.g., by using drawings (such as a beaker with a measurement scale) to represent the problem.
- 3.MD.B.4. Generate measurement data by measuring lengths using rulers marked with halves and fourths of an inch. Show the data by making a line plot, where the horizontal scale is marked off in appropriate units—whole numbers, halves, or quarters.
- 3.MD.C.5. Recognize area as an attribute of plane figures and understand concepts of area measurement.
- 3.MD.C.6. Measure areas by counting unit squares (square cm, square m, square inches, square feet, and improvised units).
- 3.MD.C.7. Relate area to the operations of multiplication and addition.
- 3.OA.B.5. Apply properties of operations as strategies to multiply and divide.
- 3.OA.D.9. Identify arithmetic problems (including patterns in the addition table or multiplication table), and explain them using properties of operations.

READING STANDARDS

- RI.3.5. Use text features and search tools (e.g., key words, sidebars, hyperlinks) to locate information relevant to a given topic efficiently.
- RI.3.7. Use information gained from illustrations (e.g., maps, photographs) and the words in a text to demonstrate understanding of the text (e.g., where, when, why, and how key events occur).
- RI.3.8. Describe the logical connection between particular sentences and paragraphs in a text (e.g., comparison, cause/effect, first/second/third in a sentence).
- RI.3.10: By the end of the year, read and comprehend informational texts, including history/social studies, science, and technical texts, at the high end of the grades 2–3 text complexity band independently and proficiently.

WRITING STANDARDS

- W.3.1. Write opinion pieces on topics or texts, supporting a point of view with reasons.
- W.3.1.A. Introduce the topic or text they are writing about, state an opinion, and create an organizational structure that lists reasons.
- W.3.1.B. Provide reasons that support the opinion.
- W.3.1.C. Use linking words and phrases (e.g., *because, therefore, since, for example*) to connect opinion and reasons.
- W.3.1.D. Provide a concluding statement or section.
- W.3.2. Write informative/explanatory texts to examine a topic and covey ideas and information clearly.
- W.3.2.B. Develop the topic with facts, definitions, and details.
- W.3.3. Write narratives to develop real or imagined experiences or events using effective technique, descriptive details, and clear event sequences.
- W.3.7. Conduct short research projects that build knowledge about a topic.
- W.3.8. Recall information from experiences or gather information from print and digital sources; take brief notes on sources and sort evidence into provided categories.

Continued

Swing Set Makeover, Grade 3

Table A2. (*continued*)

	SPEAKING AND LISTENING STANDARDS
	• SL.3.1. Engage effectively in a range of collaborative discussions (one-on-one, in groups, and teacher-led) with diverse partners on *grade 3 topics and texts,* building on others' ideas and expressing their own clearly.
	• SL.3.1.D. Explain their ideas and understanding in light of the discussion.
	• SL.3.3. Ask and answer questions about information from a speaker, offering appropriate elaboration and detail.
	• SL.3.4. Report on a topic or text, tell a story, or recount an experience with appropriate facts and relevant, descriptive details, speaking clearly at an understandable pace.
	• SL.3.6. Speak in complete sentences when appropriate to task and situations in order to provide requested detail or clarification.

Source: National Governors Association Center for Best Practices and Council of Chief State School Officers (NGAC and CCSSO). 2010. *Common core state standards.* Washington, DC: NGAC and CCSSO.

Table A3. 21st Century Skills From the Framework for 21st Century Learning

21st Century Skills	Learning Skills and Technology Tools	Teaching Strategies	Evidence of Success
INTERDISCIPLINARY THEMES	• Health and Safety • Environmental Literacy • Science • Mathematics • Engineering Design Process (EDP)	• A student-created measure of fun will be reconciled with safety considerations. • Students will explore and document the influence of position and force on motion in a variety of activities. • Student teams will use the EDP throughout the module to design and create models of components of a swing set for the culminating module challenge, the Swing Set Makeover Design Challenge.	• Students will interpret, organize, and present information from activities and research, demonstrating their understanding of force and motion and the EDP as well as of the need for safety considerations in swing set design.
LEARNING AND INNOVATION SKILLS	• Creativity and Innovation • Critical Thinking and Problem Solving • Communication and Collaboration	• Teach and facilitate creativity by encouraging students to think outside the box to solve problems. • Facilitate group work and instruct students on internet search procedures and strategies.	• Students will record their EDP thinking in multiple formats and work with their teams to share and improve on their swing set models (sketches). • Students will document and support their ideas with evidence recorded in their STEM Research Notebooks. • Design teams will interact in activities that reinforce the importance of communication and collaboration.

Continued

Table A3. (*continued*)

21st Century Skills	Learning Skills and Technology Tools	Teaching Strategies	Evidence of Success
INFORMATION, MEDIA, AND TECHNOLOGY SKILLS	• Information Literacy • Media Literacy • ICT Literacy	• Students use technology to conduct research to gain an understanding of swing set design and safety considerations and an appreciation for different types of public parks. • Students use technology to share experiences and knowledge in a blog.	• Student blogs includes evidence from research and classroom inquiry experiences.
LIFE AND CAREER SKILLS	• Flexibility and Adaptability • Initiative and Self-Direction • Social and Cross-Cultural Skills • Productivity and Accountability • Leadership and Responsibility	• Scaffold sketch-and-model through a series of inquiry activities and topical research projects. • Use EDP to encourage flexibility (through redesign), time management, and goal setting in structured group work. • Provide guidelines and practice opportunities for students to share, emphasizing professional standards of behavior and inclusivity of all team members.	• Team projects are completed on time with evidence of participation by all team members. • Teams' presentations include appropriate language and vocabulary. • Students are able to respond to questions regarding their design process and teamwork.

Source: Partnership for 21st Century Learning. 2015. Framework for 21st Century Learning. *www.p21.org/our-work/p21-framework.*

Table A4. English Language Development (ELD) Standards

ELD STANDARD 1: SOCIAL AND INSTRUCTIONAL LANGUAGE

English language learners communicate for Social and Instructional purposes within the school setting.

ELD STANDARD 2: THE LANGUAGE OF LANGUAGE ARTS

English language learners communicate information, ideas, and concepts necessary for academic success in the content area of Language Arts.

ELD STANDARD 3: THE LANGUAGE OF MATHEMATICS

English language learners communicate information, ideas, and concepts necessary for academic success in the content area of Mathematics.

ELD STANDARD 4: THE LANGUAGE OF SCIENCE

English language learners communicate information, ideas, and concepts necessary for academic success in the content area of Science.

ELD STANDARD 5: THE LANGUAGE OF SOCIAL STUDIES

English language learners communicate information, ideas, and concepts necessary for academic success in the content area of Social Studies.

Source: WIDA. 2012. 2012 amplification of the English language development standards: Kindergarten–grade 12. *https://wida.wisc.edu/teach/standards/eld.*

INDEX

Page numbers printed in **boldface type** indicate tables, figures, or handouts.